NAVIGATING
through
DATA ANALYSIS
in
GRADES 9–12

Gail Burrill
Christine A. Franklin
Landy Godbold
Linda J. Young

Johnny W. Lott
Grades 9–12 Editor

Peggy A. House
Navigations Series Editor

NCTM

NATIONAL COUNCIL OF
TEACHERS OF MATHEMATICS

NAVIGATIONS SERIES

TABLE OF CONTENTS

CONTENTS OF THE CD-ROM

Introduction

Table of Standards and Expectations, Data Analysis and Probability, Pre-K–12

Applets and Activities

Random Number Generator
Random Rectangles
Discrimination or Not
Lowering Cholesterol—Counting Successes
Lowering Cholesterol—Mean Improvement
Populations and Samples

Blackline Masters and Templates

Readings and Supplemental Materials

Beyond Exploratory Data Analysis: The Randomization Test
    Peter Barbella, Lorraine Denby, and James M. Landwehr
    Mathematics Teacher

Investigating Distributions of Sample Means on the Graphing Calculator
    Gloria B. Barrett
    Mathematics Teacher

Analyzing Data from the Olympic Games for Trends and Inferences
    Richard T. Edgerton
    Mathematics Teacher

Spreadsheets: Powerful Tools for Probability Simulations
    Edwin McClintock and Zhonghong Jiang
    Mathematics Teacher

About This Book

Today's world is a very different place from the world of thirty years ago. We use magnetic resonance imaging (MRI) to diagnose medical problems, computer-assisted design (CAD) for engineering and architectural plans, spreadsheets to analyze investments, simulations to plan travel in space, and the Internet for instant communication across cities and nations. These are just a few of the information advances of our time. Information has transformed the way people live and the way they do business. The ability to make sound decisions on the basis of information is an increasingly indispensable skill. What does this signify for the mathematical content that students need to become productive citizens and workers as they move out into the world? The data analysis strand in *Principles and Standards for School Mathematics* (National Council of Teachers of Mathematics [NCTM] 2000) addresses this question by describing the important ideas that are central to understanding how to process information and use it wisely. *Navigating through Data Analysis in Grades 9–12* is about making good decisions on the basis of well-designed studies and using ideas about probability and randomness as tools in decision making.

Data analysis can be thought of as consisting of two parts: exploration and inference (Cobb 1997). *Navigating through Data Analysis in Grades 9–12* builds students' intuition by enabling them to explore several real problems that can help them begin to use statistical reasoning about data. *Principles and Standards for School Mathematics* suggests that data analysis in grades 9–12 should build on the foundations established from prekindergarten through grade 8. In the early grades, students should have conducted explorations and learned techniques for graphing data and calculating and interpreting summary statistics. They probably began those early experiences by learning to formulate "questions that can be addressed with data and collect[ing], organiz[ing], and display[ing] relevant data to answer them" (NCTM 2000, p. 324). In high school, the curriculum for all students should focus more attention on careful statistical reasoning. Students should be expected to find ways to generalize about patterns while using the exploratory skills learned in earlier grades, but now the emphasis ought to be on drawing conclusions.

Data from which high school students can reason can be collected through *observational studies*, *sample surveys*, or *experiments*. Data may be *categorical* or *numerical*. Categorical data represent characteristics of individuals or objects in a population by placing these individuals or objects in groups (or categories), such as male or female, or red or blue. Numerical data are measurements or counts taken on each object or individual in a population. Numerical data may be either *discrete*—for example, the number of defective television sets in a warehouse—or *continuous*—for instance, the time required to run a mile. The material in this book centers on the use of simulated sampling distributions, summary statistics, and graphs to reason from data. In the activities that the book provides, students evaluate published reports that are based on data and consider the role of chance and variation by examining the design of the studies, the appropriateness of their analyses of data, and their conclusions.

Variability

Individuals, objects, and repeated measurements of the same object vary. Statistics is the science of reasoning from data in the presence of this variability. Thus, a central theme of *Navigating through Data Analysis in Grades 9–12* is the role that variability plays in the claims that people make about the results of an experiment, sample survey, or observational study. The role of *randomization*, or chance, is fundamental to the idea of using observations from a few to draw conclusions about a larger group containing the few.

Chance behavior, although unpredictable in the short run, has a regular and predictable pattern in the long run. You cannot predict whether you will get a head or a tail on one toss of a fair coin, but if you toss the coin 100 times, you can expect to get heads approximately 50 times. Some outcomes are likely, whereas others, such as the coin's happening to land on its edge, are very unlikely, although still possible. Getting heads 45 times in 100 tosses would be credible, but if heads turned up only 5 times, you might raise questions about the coin. Statistical inference, which makes use of this long-term predictability, is based on an explicit chance model for the data (Freedman, Pisani, and Purves 1998).

Data analysis generally focuses on using data from a sample or an experiment to learn more about a population. A *population* is the entire group of objects about which you want information—for example, all ninth graders with a certain medical condition. A *sample* is the part of the population from which you are actually collecting information—some subset of ninth graders with the medical condition, for instance. A *parameter* is a numerical summary that describes a particular characteristic of the whole population, such as the mean age of all the fish in a lake or the proportion of all the people in the United States who agree with a given statement. A *statistic* is a numerical summary that is computed from sample data to estimate an unknown population parameter.

The statistic might be the mean age of a sample of fish randomly drawn from a lake or the proportion of a sample of people in the United States who agree with a given statement. The value of a statistic is known for a given sample but can change from sample to sample. The values of a statistic for repeated samples of the same size from a given population generate a *sampling distribution*. Sampling distributions answer the question "How would the statistic behave if the process were repeated many, many times under the same conditions?" In other words, what is the predictable pattern of the *variability* in the sample statistic? In statistics, *simulations* are designed to use chance to model a situation or process. In this book, students use simulation activities to explore the variability of sample statistics from a population by constructing simulated sampling distributions.

Activities in this book assume that students can identify and describe the shape, center, spread, and outlying values of a distribution and that they also know how to make and interpret statistical plots, including bar

The notion of deliberately using chance in producing data, popularized by R. A. Fisher in 1919, is one of the most important ideas in statistics (Moore 2000, p. 164). The increasing power of technology can make this idea very accessible to students today.

The activity **Populations and Samples,** *which appears on the accompanying CD-ROM, can be used to give students an introductory experience in identifying populations and samples from those populations.*

charts, histograms, scatterplots, box plots, line plots, and dot plots. The center of a distribution is usually measured by the *mean*—the arithmetic mean, or average, of the values—or by the *median*—the point that marks the middle value in an ordered set of values or, by convention, the mean of the two middle values. Measures of spread include the *range*—the distance between the maximum and minimum values—and the *interquartile range*—the distance covered by the middle 50 percent of the values. Another measure of spread is the *mean absolute deviation*—an "average" distance of values from the mean. The most common measure of spread, however, is the *standard deviation*. The standard deviation can be thought of as the size of a "typical" deviation between the mean and a value. Students may need experience in using one of these measures of spread for several of the exercises. Table 0.1 summarizes measures of center and spread.

Table 0.1
Measures of Center and Spread

Measure of Center	Measure of Spread
Mean	Standard deviation
	Mean absolute deviation
	Range
Median	Interquartile range

See Measures of Variability on the accompanying CD-ROM for a more detailed discussion of standard deviation.

Mary Bud Rowe
think time—
article?

Teaching Data Analysis

As you present each activity in the book, pose questions to your students, allow them time to think about possible responses, and ask them to share their ideas. To learn "with understanding, actively building new knowledge from experience and prior knowledge" (NCTM 2000, p. 11), students must be involved in developing ideas and exploring conjectures. The following principles serve to guide and foster this kind of learning:

Students should—

* understand the need for and be able to formulate well-defined questions;
* study the role that randomization plays in producing data;
* identify assumptions;
* have hands-on experiences in producing data;
* begin with a picture of the data; and
* be involved in organizing and analyzing the data.

In addition, students should—

* explore and experiment before using formal algorithms;
* develop a conceptual understanding of statistical concepts; and
* focus on the "big ideas" instead of the rules.

For example, students need to generate their own sets of data and create simulated sampling distributions to understand variation. Through experience, they will begin to understand that tossing a fair coin six

The CD-ROM includes readings from the *Mathematics Teacher* on exploratory data analysis, and readers can also consult *Navigating through Data Analysis in Grades 6–8* (Bright et al. 2003) for additional information about statistical plots that most students have already encountered by the time they reach grades 9–12.

times can produce six heads and will see how a sampling distribution can be simulated. From such experiences, they can build a sense of how to begin to reason about data.

Although hands-on activities are important for students who are at this stage, technology is crucial to their development of meaningful insights. Technology allows students to change one or more observations and instantly see the consequences. It also enables them to make comparisons of the numerical summaries for a data set with parameters of the distribution. To internalize what distributions are, however, students should begin with techniques that allow them to work "by hand," such as tossing coins and dice to simulate simple situations. In chapter 1, the activities direct students to use calculators or computers to generate random numbers and do calculations. In chapter 2, they use decks of cards to model a scenario called Discrimination or Not? Once students understand basic concepts, software such as Fathom (KCP Technologies 2000) can be useful for exploring ideas and building models.

As you teach, relate the study of data analysis to other areas of mathematics. The language of algebra is also the language of statistics. Certain technical skills are prerequisites for understanding and efficiently using the language of both areas. The mathematics curriculum should build an understanding of the formulas for mean and standard deviation just as it builds understanding of formulas for surface area and volume. Some statistical phenomena may be represented in the same way that functions are—in tables, graphs, and symbols. As noted in *Navigating through Algebra in Grades 9–12* (Burke et al. 2000), a *variable* is a named quantity in both statistics and algebra and can have many values in both. Just as algebraic functions have important features, such as maxima and slopes, distributions have their own important features, including centers and spreads. And just as the shape of an algebraic curve indicates something about the nature of the relationship between the domain and the range, the shape of a distribution indicates something about the nature of the underlying variable. Both statistical and mathematical reasoning depend on clearly stated assumptions.

The Sequence of Chapters

A general introduction by series editor Peggy A. House opens all six of the Navigations books that elaborate NCTM's combined Data Analysis and Probability Standard. This introduction gives a broad overview of the appropriate development of the concepts and skills related to data analysis and probability in students from prekindergarten through grade 12. This book's own exploration of data analysis in the high school years begins in chapter 1. Activities in this chapter enable students to investigate simple random samples, in which every subset of a given size from the population has an equal chance of being selected. Students also explore how sample size and different sampling techniques affect the variability of a statistic.

Chapter 2 gives students an opportunity to explore an actual case study involving potential discrimination and, through simulation, helps them decide whether the data support a claim of discrimination against women. The fundamental question is "How likely would it be that something as unusual as the observed outcome would occur by chance if there were no discrimination?"

Chapter 3 allows students to investigate a case study involving cholesterol levels, exploring the relationship between two variables and considering a variety of approaches for analyzing the data. The use of simulation enables students to understand how likely it would be that the observed number of volunteers would show improvement by chance if the treatment had no effect.

Chapter 4 builds an understanding of observational and experimental studies and explores the features of well-designed studies. Chapter 5 suggests several activities that you might assign for assessment purposes, to see whether your students can put statistical thinking into practice.

Each chapter is built on a set of activities that you can use with your students or for your own professional development. The activities include lists of necessary materials and provide activity sheets in the form of reproducible blackline masters, which can also be printed from the CD-ROM that accompanies the book. Interactive applets on the CD-ROM provide additional hands-on opportunities. Extensions, supporting materials, and a fuller description of how some of the activities relate to probability and formal statistical inference are also available on the accompanying CD-ROM, together with supplemental readings for professional development.

Three different icons appear in the book, as shown in the key. One alerts readers to material quoted from *Principles and Standards for School Mathematics*, another points them to supplementary materials on the CD-ROM that accompanies the book, and a third signals the blackline masters and indicates their locations in the appendix.

Navigating through Data Analysis in Grades 9–12 offers new material and thinking for high school students and emphasizes the relevance of data analysis in today's world. The purpose of the book is to help you lay the foundations for your students to understand and use data to make intelligent, justifiable decisions in the face of the enormous amount of information available all around them. As you read the book and work through the activities with your students, invite them to consider how the materials relate to what is important in decisions that they make or that are made for them in their lives. Check newspapers and other media for current topics that you and your students might subject to the kind of analysis that *Navigating through Data Analysis in Grades 9–12* illustrates. We hope that in all of these efforts, you will enjoy—and help your students enjoy—the sense of clarity and control that being able to analyze data gives!

Key to Icons

Principles and Standards   CD-ROM   Blackline Master

We gratefully acknowledge the major contributions of the reviewers of this book:

Tim Erickson
Patrick Hopfensperger
Jon Graham
Richard Scheaffer
Brian Steele

—*The authors*

Any collaborative work requires a monumental effort. This one called for a truly herculean effort. Special contributions were made by Gail Burrill, who helped assemble the team and led early editing, and Anita Draper, NCTM staff editor, who persevered.

—*J. W. Lott*

NAVIGATING *through* DATA ANALYSIS

Introduction

real life?

The Data Analysis and Probability Standard in *Principles and Standards for School Mathematics* (NCTM 2000) is an affirmation of a fundamental goal of the mathematics curriculum: to develop critical thinking and sound judgment based on data. These skills are essential not only for a select few but for every informed citizen and consumer. Staggering amounts of information confront us in almost every aspect of contemporary life, and being able to ask good questions, use data wisely, evaluate claims that are based on data, and formulate defensible conclusions in the face of uncertainty have become basic skills in our information age.

In working with data, students encounter and apply ideas that connect directly with those in the other strands of the mathematics curriculum as well as with the mathematical ideas that they regularly meet in other school subjects and in daily life. They can see the relationship between the ideas involved in gathering and interpreting data and those addressed in the other Content Standards—Number and Operations, Algebra, Measurement, and Geometry—as well as in the Process Standards— Reasoning and Proof, Representation, Communication, Connections, and Problem Solving. In the Navigations series, the *Navigating through Data Analysis and Probability* books elaborate the vision of the Data Analysis and Probability Standard outlined in *Principles and Standards*. These books show teachers how to introduce important statistical and probabilistic concepts, how the concepts grow, what to expect students to be able to do and understand during and at the end of each grade band, and how to assess what they know. The books also introduce representative instructional activities that help translate the vision of *Principles and Standards* into classroom practice and student learning.

Fundamental Components of Statistical and Probabilistic Thinking

Principles and Standards sets the Data Analysis and Probability Standard in a developmental context. It envisions teachers as engaging students from a very young age in working directly with data, and it sees this work as continuing, deepening and growing in sophistication and complexity as the students move through school. The expectation is that all students, in an age-appropriate manner, will learn to—

- formulate questions that can be addressed with data and collect, organize, and display relevant data to answer them;
- select and use appropriate statistical methods to analyze data;
- develop and evaluate inferences and predictions that are based on data; and
- understand and apply basic concepts of probability.

Formulating questions that can be addressed with data and collecting, organizing, and displaying relevant data to answer them

No one who has spent any time at all with young children will doubt that they are full of questions. Teachers of young children have many opportunities to nurture their students' innate curiosity while demonstrating to them that they themselves can gather information to answer some of their questions.

At first, children are primarily interested in themselves and their immediate surroundings, and their questions center on such matters as "How many children in our class ride the school bus?" or "What are our favorite flavors of ice cream?" Initially, they may use physical objects to display the answers to their questions, such as a shoe taken from each student and placed appropriately on a graph labeled "The Kinds of Shoes Worn in Kindergarten." Later, they learn other methods of representation using pictures, index cards, sticky notes, or tallies. As children move through the primary grades, their interests expand outward to their surroundings, and their questions become more complex and sophisticated. As that happens, the amount of collectible data grows, and the task of keeping track of the data becomes more challenging. Students then begin to learn the importance of framing good questions and planning carefully how to gather and display their data, and they discover that organizing and ordering data will help uncover many of the answers that they seek. However, learning to refine their questions, planning effective ways to collect data, and deciding on the best ways to organize and display data are skills that children develop only through repeated experiences, frequent discussions, and skillful guidance from their teachers. By good fortune, the primary grades afford many opportunities—often in conjunction with lessons on counting, measurement, numbers, patterns, or other school subjects— for children to pose interesting questions and develop ways of collecting data that will help them formulate answers.

As students move into the upper elementary grades, they will continue to ask questions about themselves and their environment, but their questions will begin to extend to their school or the community or the world beyond. Sometimes, they will collect their own data; at other times, they will use existing data sets from a variety of sources. In either case, they should learn to exercise care in framing their questions and determining what data to collect and when and how to collect them. They should also learn to recognize differences among data-gathering techniques, including observation, measurement, experimentation, and surveying, and they should investigate how the form of the questions that they seek to answer helps determine what data-gathering approaches are appropriate. During these grades, students learn additional ways of representing data. Tables, line plots, bar graphs, and line graphs come into play, and students develop skill in reading, interpreting, and making various representations of data. By examining, comparing, and discussing many examples of data sets and their representations, students will gain important understanding of such matters as the difference between categorical and numerical data, the need to select appropriate scales for the axes of graphs, and the advantages of different data displays for highlighting different aspects of the same data.

During middle school, students move beyond asking and answering the questions about a single population that are common in the earlier years. Instead, they begin posing questions about relationships among several populations or samples or between two variables within a single population. In grades 6–8, students can ask questions that are more complex, such as "Which brand of laundry detergent is the best buy?" or "What effect does light [or water or a particular nutrient] have on the growth of a tomato plant?" They can design experiments that will allow them to collect data to answer their questions, learning in the process the importance of identifying relevant data, controlling variables, and choosing a sample when it is impossible to collect data on every case. In these middle school years, students learn additional ways of representing data, such as with histograms, box plots, or relative-frequency bar graphs, and they investigate how such displays can help them compare sets of data from two or more populations or samples.

By the time students reach high school, they should have had sufficient experience with gathering data to enable them to focus more precisely on such questions of design as whether survey questions are unambiguous, what strategies are optimal for drawing samples, and how randomization can reduce bias in studies. In grades 9–12, students should be expected to design and evaluate surveys, observational studies, and experiments of their own as well as to critique studies reported by others, determining if they are well designed and if the inferences drawn from them are defensible.

Selecting and using appropriate statistical methods to analyze data

Teachers of even very young children should help their students reflect on the displays that they make of the data that they have gathered. Students should always thoughtfully examine their representations to determine what information they convey. Teachers can prompt young children to derive information from data displays through questions like "Do more children in our class prefer vanilla ice cream, or do more prefer chocolate ice cream?" As children try to interpret their work, they come to realize that data must be ordered and organized to convey answers to their questions. They see how information derived from data, such as their ice cream preferences, can be useful—in deciding, for example, how much of particular flavors to buy for a class party. In the primary grades, children ordinarily gather data about whole groups—frequently their own class—but they are mainly interested in individual data entries, such as the marks that represent their own ice cream choices. Nevertheless, as children move through the years from prekindergarten to grade 2, they can be expected to begin questioning the appropriateness of statements that are based on data. For example, they may express doubts about such a statement as "Most second graders take ballet lessons" if they learn that only girls were asked if they go to dancing school. They should also begin to recognize that conclusions drawn about one population may not apply to another. They may discover, for instance, that bubble gum and licorice are popular ice cream flavors among their fellow first graders but suspect that this might not necessarily be the case among their parents.

In contrast with younger children, who focus on individual, often personal, aspects of data sets, students in grades 3–5 can and should be guided to see data sets as wholes, to describe whole sets, and to compare one set with another. Students learn to do this by examining different sets' characteristics—checking, for example, values for which data are concentrated or clustered, values for which there are no data, or values for which data are unusually large or small (*outliers*). Students in these grades should also describe the "shape" of a whole data set, observing how the data spread out to give the set its *range*, and finding that range's center. In grades 3–5, the center of interest is in fact very often a measure of a data set's center—the *median* or, in some cases, the *mode*. In the process of learning to focus on sets of data rather than on individual entries, students should start to develop an understanding of how to select *typical* or *average* (*mean*) values to represent the sets. In examining similarities and differences between two sets, they should explore what the means and the ranges tell about the data. By using standard terms in their discussions, students in grades 3–5 should be building a precise vocabulary for describing the characteristics of the data that they are studying.

By grade 5, students may begin to explore the concept of the mean as a balance point in an informal way, but a formal understanding of the mean and its use in describing data sets does not become important until grades 6–8. By this time, just being able to compute the mean is no longer enough. Students need ample opportunities to develop a fundamental conceptual understanding—for example, by comparing the usefulness and appropriateness of the mean, the median, and the mode as ways of describing data sets in different contexts. In middle school, students should also explore questions that are more probing, such as "What impact does the spread of a distribution have on the value of the mean [or the median]?" Or "What effect does changing one data value [or more than one] have on different measures of center—the mean, the median, and the mode?" Technology, including spreadsheet software, calculators, and graphing software, becomes an important tool in grades 6–8, enabling students to manipulate and control data while they investigate how changes in certain values affect the mean, the median, or the distribution of a set of data. Students in grades 6–8 should also study important characteristics of data sets, such as *symmetry*, *skewness*, and *interquartile range*, and should investigate different types of data displays to discover how a particular representation makes such characteristics more or less apparent.

As these students move on into grades 9–12, they should grow in their ability to construct an appropriate representation for a set of univariate data, describe its shape, and calculate summary statistics. In addition, high school students should study linear transformations of univariate data, investigating, for example, what happens if a constant is added to each data value or if each value is multiplied by a common factor. They should also learn to display and interpret bivariate data and recognize what representations are appropriate under particular conditions. In situations where one variable is categorical—for example, gender—and the other is numerical—a measurement of height, for instance—students might use appropriately paired box plots or histograms to compare the heights of males and females in a given group. By contrast, students who are presented with bivariate numerical data—for example, measurements of height and arm span—might use a scatterplot to represent their data, and they should be able to describe the shape of the scatterplot and use it to analyze the relationship between the two lengths measured—height and arm span. Types of analyses expected of high school students include finding functions that approximate or "fit" a scatterplot, discussing different ways to define "best fit," and comparing several functions to determine which is the best fit for a particular data set. Students should also develop an understanding of new concepts, including *regression*, *regression line*, *correlation*, and *correlation coefficient*. They should be able to explain what each means and should understand clearly that a correlation is not the same as a causal relationship. In grades 9–12, technology that allows users to plot, move, and compare possible regression lines can help students develop a conceptual understanding of residuals and regression lines and can enable them to compute the equation of their selected line of best fit.

Developing and evaluating inferences and predictions that are based on data

Observing, measuring, or surveying every individual in a population is an appropriate way of gathering data to answer selected questions. Such "census data" is all that we expect from very young children, and teachers in the primary grades should be content when their students confine their data gathering and interpretation to their own class or another small group. But as children mature, they begin to understand that a principal reason for gathering and analyzing data is to make inferences and predictions that apply beyond immediately available data sets. To do that requires sampling and other more advanced statistical techniques.

Teachers of young children lay a foundation for later work with inference and prediction when they ask their students whether they think that another group of students would get the same answers from data that they did. After discussing the results of a survey to determine their favorite books, for example, children in one first-grade class might conclude that their peers in the school's other first-grade class would get similar results but that the fourth graders' results might be quite different. The first graders could speculate about why this might be so.

As students move into grades 3–5, they should be expected to expand their ability to draw conclusions, make predictions, and develop arguments based on data. As they gain experience, they should begin to understand how the data that they collect in their own class or school might or might not be representative of a larger population of students. They can begin to compare data from different samples, such as several fifth-grade classes in their own school or other schools in their town or state. They can also begin to explore whether or not samples are representative of the population and identify factors that might affect representativeness. For example, they could consider a question like "Would a survey of children's favorite winter sports get similar results for samples drawn from Colorado, Hawaii, Texas, and Ontario?" Students in the upper grades should also discuss differences in what data from different samples show and factors that might account for the observed results, and they can start developing hypotheses and designing investigations to test their predictions.

It is in the middle grades, however, that students learn to address matters of greater complexity, such as the relationship between two variables in a given population or sample, or the relationships among several populations or samples. Two concepts that are emphasized in grades 6–8 are *linearity* and *proportionality*, both of which are important in developing students' ability to interpret and draw inferences from data. By using scatterplots to represent paired data from a sample—for example, the height and stride length of middle schoolers—students might observe whether the points of the scatterplot approximate a line, and if so, they can attempt to draw the line to fit the data. Using the slope of that line, students can make conjectures about a relationship between height and stride length. Furthermore, they might decide to compare a scatterplot for middle school boys with one for middle school girls to determine if a similar ratio applies for both groups. Or they might draw box plots or relative-frequency histograms to represent data on the heights of samples of middle school boys and high school boys to investigate the variability in height of boys of different ages. With the help of graphing technology, students can examine many data sets and learn to differentiate between linear and nonlinear relationships, as well as to recognize data sets that exhibit no relationship at all. Whenever possible, they should attempt to describe observed relationships mathematically and discuss whether the conjectures that they draw from the sample data might apply to a larger population. From such discussions, students can plan additional investigations to test their conjectures.

As students progress to and through grades 9–12, they can use their growing ability to represent data with regression lines and other mathematical models to make and test predictions. In doing so, they learn that inferences about a population depend on the nature of the samples, and concepts such as *randomness, sampling distribution*, and *margin of error* become important. Students will need firsthand experience with many different statistical examples to develop a deep understanding of the powerful ideas of inference and prediction. Often that experience will come through simulations that enable students to perform hands-on experiments while developing a more intuitive understanding of the relationship between characteristics of a sample and the corresponding characteristics of the population from which the sample was drawn. Equipped with the concepts learned through simulations, students can then apply their understanding by analyzing statistical inferences and critiquing reports of data gathered in various contexts, such as product testing, workplace monitoring, or political forecasting.

Understanding and applying basic concepts of probability

Probability is connected to all mathematics from number to geometry. It has an especially close connection to data collection and analysis. Although students are not developmentally ready to study probability in a formal way until much later in the curriculum, they should begin to lay the foundation for that study in the years from prekindergarten to grade 2. For children in these early years, this means informally considering ideas of likelihood and chance, often by thinking about such questions as "Will it be warm tomorrow?" and realizing that the answer may depend on particular conditions, such as where they live or what month it is. Young children also recognize that some things are sure to happen whereas others are impossible, and they begin to develop notions of *more likely* and *less likely* in various everyday contexts. In addition, most children have experience with common devices of chance used in games, such as spinners and dice. Through hands-on experience, they become aware that certain numbers are harder than others to get with two dice and that the pointer on some spinners lands on certain colors more often than on others.

In grades 3–5, students can begin to think about probability as a measurement of the likelihood of an event, and they can translate their earlier ideas of *certain, likely, unlikely*, or *impossible* into quantitative representations using 1, 0, and common fractions. They should also think about events that are neither certain nor impossible, such as getting a 6 on the next roll of a die. They should begin to understand that although they cannot know for certain what will happen in such a case, they can associate with the outcome a fraction that represents the frequency with which they could expect it to occur in many similar situations. They can also use data that they collect to estimate probability—for example, they can use the results of a survey of students' footwear to predict whether the next student to get off the school bus will be wearing brown shoes.

Students in grades 6–8 should have frequent opportunities to relate their growing understanding of proportionality to simple probabilistic situations from which they can develop notions of chance. As they refine their understanding of the chance, or likelihood, that a certain event will occur, they develop a corresponding sense of the likelihood that it will not occur, and from this awareness emerge notions of complementary events, mutually exclusive events, and the relationship between the probability of an event and the probability of its complement. Students should also investigate simple compound events and use tree diagrams, organized lists, or similar descriptive methods to determine probabilities in such situations. Developing students' understanding of important concepts of probability—not merely their ability to compute probabilities—should be the teacher's aim. Ample experience is important, both with hands-on experiments that generate empirical data and with computer simulations that produce large data samples. Students should then apply their understanding of probability and proportionality to make and test conjectures about various chance events, and they should use simulations to help them explore probabilistic situations.

Concepts of probability become increasingly sophisticated during grades 9–12 as students develop an understanding of such important ideas as *sample space, probability distribution, conditional probability, dependent* and *independent events,* and *expected value.* High school students should use simulations to construct probability distributions for sample spaces and apply their results to predict the likelihood of events. They should also learn to compute expected values and apply their knowledge to determine the fairness of a game. Teachers can reasonably expect students at this level to describe and use a sample space to answer questions about conditional probability. The solid understanding of basic ideas of probability that students should be developing in high school requires that teachers show them how probability relates to other topics in mathematics, such as counting techniques, the binomial theorem, and the relationships between functions and the area under their graphs.

Developing a Data Analysis and Probability Curriculum

Principles and Standards reminds us that a curriculum that fosters the development of statistical and probabilistic thinking must be coherent, focused, and well articulated—not merely a collection of lessons or activities devoted to diverse topics in data analysis and probability. Teachers should introduce rudimentary ideas of data and chance deliberately and purposefully in the early years, deepening and expanding their students' understanding of them through frequent experiences and applications as students progress through the curriculum. Students must be continually challenged to learn and apply increasingly sophisticated statistical and probabilistic thinking and to solve problems in a variety of school, home, and real-life settings.

The six *Navigating through Data Analysis and Probability* books make no attempt to present a complete, detailed data analysis and probability curriculum. However, taken together, these books illustrate how selected "big ideas" behind the Data Analysis and Probability Standard develop this strand of the mathematics curriculum from prekindergarten through grade 12. Many of the concepts about data analysis and probability that the books present are closely tied to topics in algebra, geometry, number, and measurement. As a result, the accompanying activities, which have been especially designed to put the Data Analysis

and Probability Standard into practice in the classroom, can also reinforce and enhance students' understanding of mathematics in the other strands of the curriculum, and vice versa.

Because the methods and ideas of data analysis and probability are indispensable components of mathematical literacy in contemporary life, this strand of the curriculum is central to the vision of mathematics education set forth in *Principles and Standards for School Mathematics*. Accordingly, the *Navigating through Data Analysis and Probability* books are offered to educators as guides for setting successful courses for the implementation of this important Standard.

NAVIGATING *through* DATA ANALYSIS

Chapter 1
Decision Making, Variability, and Sampling

H. G. Wells once wrote that "statistical thinking will one day be as necessary for efficient citizenship as the ability to read and write" (Huff 1954). That this has become true today is recognized in *Principles and Standards for School Mathematics* (NCTM 2000), where data analysis and probability are paired as one of the Content Standards.

The usefulness of statistical skills is clear in many situations. We sometimes suspect that events haven't happened the way they did by pure chance, and we would like to confirm—with statistical evidence, if possible—our suspicions that someone or something "stacked the deck." Chapter 2 presents a situation of this type from an actual study of possible discrimination against women in the workplace. In this study, 48 male supervisors each reviewed a personnel folder of an employee supposedly being considered for promotion. Twenty-four copies of the folder were labeled "male," and 24 were labeled "female." Depending on the actual numbers of male and female candidates recommended for promotion, could we determine whether discrimination against women played a role in the supervisors' recommendations? Activities in chapter 2 invite students to explore this question.

Using statistics to answer such a question involves understanding fundamental concepts of data analysis. A primary consideration is how the data were obtained. In most instances, it is not possible to gather data from the entire population under consideration. Thus, an analysis usually relies on a sample taken from this population. Taking samples raises concerns about the method of selection and the ability to use sample results to reason about a population. The goal of sampling is to be able to make reasonable statements about the population from the sample data.

Time, cost, efficiency, and the situation itself often necessitate using samples rather than taking a *census*, which involves collecting information from all units or members belonging to the entire population that is being considered. For example, eating all the soup in a kettle to see if it is well seasoned defeats the purpose of making the soup; checking a sample should suffice. However, those who sample—using a part or parts to represent the whole—must remember that different samples from the same population will differ. In other words, samples vary. For data analysis, a sample should be gathered in such a way that the statistic computed from the sample varies from sample to sample in a predictable way.

The "big ideas" in this chapter deal with sampling and the variability that is present in a sampling distribution. The focus is on exploring simple random sampling, sampling distributions, the effect of sample size on variability, and alternative sampling methods. The activities in this chapter use a single population—a set of 100 random rectangles (Barbella, Kepner, and Scheaffer 1994)—to explore different techniques for sampling and to consider how these techniques are related to the larger question of variability. This population of rectangles appears in figure 1.1 and on the activity sheet "Random Rectangles."

The first activity, Sampling Rectangles, introduces simple random sampling. In this process, each possible sample of a specified size—here a five-member subset of the given population—has an equal chance of being selected. The activity enables students to compare the results of determining samples subjectively, by individual choices of "typical" rectangles, with the results of obtaining samples through a random selection process.

The second activity, Sample Size, focuses on the size of a sample and its effect on the variability of a statistic, and the third activity, Sampling Methods, considers other sampling methods that are often used in real situations.

The activity Sampling Rectangles begins with students working individually before working together as a class; the other activities using the 100 rectangles can be done in the same way, or, alternatively, students can work entirely in groups or as a whole class.

Sampling Rectangles

Goals

- Understand what is meant by a *simple random sample*
- Recognize the *variability* in subjective and random sampling techniques and understand how to measure it
- Recognize how *bias* can occur in the sampling process

Materials and Equipment

- A copy of the activity pages for each student
- A copy of the activity sheet "Random Rectangles" for each student
- (For the teacher) a transparency of the activity sheet "Distribution of Sample Mean Areas of Rectangles," plus a blank transparency
- Graphing tools to create box plots, dot plots, or histograms and to calculate simple statistics
- The Random Number Generator applet

pp. 94, 95–96, 97

Discussion

Figure 1.1 shows the population of rectangles that students encounter on the activity sheet "Random Rectangles," which students will use with all the activities in chapter 1. Each small square in the figure represents an area of 1 square unit. Distribute a copy of "Random Rectangles" to each student, along with copies of the "Sampling Rectangles" activity pages. The activity directs students to select five rectangles that they think are "typical" of the group of rectangles on the page. Give the students a limited time—maybe 15 seconds—to choose their five typical rectangles and to record the areas.

Next, students find the mean area of the rectangles that they have chosen as typical. Then the students choose five more rectangles, this time by generating random numbers between 1 and 100 and selecting the rectangles that correspond to these numbers. (The Random Number Generator applet on the CD-ROM can facilitate this work). They record their data for this "random" sample of rectangles, again calculating the sample mean for the five rectangles' areas.

Now the students are ready to pool their results (see item 3 on the blackline master "Sampling Rectangles"). Help the students work as a class to make dot plots of the two distributions—one for the sample means of the areas of the subjectively chosen rectangles and the other for the sample means of the areas of the randomly chosen rectangles. Display the simulated sampling distributions on an overhead transparency (the activity sheet "Distribution of Sample Mean Areas of Rectangles" provides a template). Item 4 on the blackline master asks students to record a measure of center and a measure of spread for the simulated sampling distribution of the mean areas of random samples of size 5. Students may also use measures of center and spread for the two distributions to compare the two methods of choosing a sample of rectangles.

The activity Sampling Rectangles can help students begin to see the difference between a sample chosen by using one's judgment and one chosen at random.

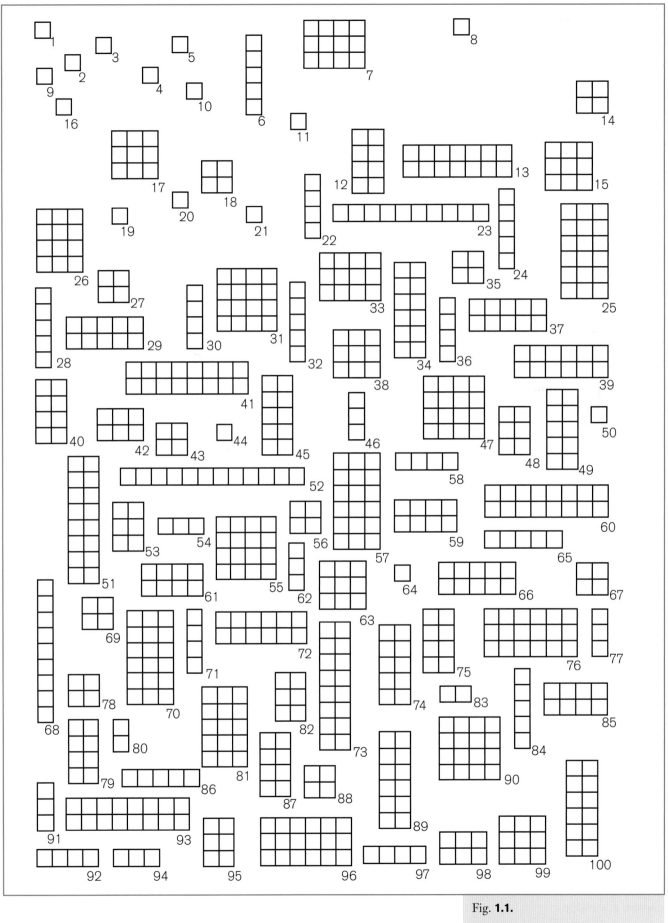

Fig. **1.1.**

Random rectangles

Randomization

People often think that they can make good judgments about what is typical. The activity Sampling Rectangles focuses on what randomization brings to the sampling process. *Random* does not mean *accidental.* A group consisting of whoever is in the school hallway or at the grocery store does not constitute a random sample for a study but can compose what is called a *convenience sample.* Convenience samples may be different in important ways from the population under consideration. The differences affect the sample's usefulness to statisticians in making predictions about the larger population. For example, to find the degree of public support for building a new baseball stadium, would it be reasonable to ask only the people at a game? Is a shopping mall a good place to ask people about a political issue? Another sampling method that is often used is compiling *voluntary* responses—for example, reports on the effects of exercise from those who offer to discuss the topic, or views on politics from those who call in to a TV talk show. Such volunteers may not be typical of the population under consideration, and conclusions drawn from such samples may not be transferable to this population. Voluntary and convenience samples may be biased; that is, the results may systematically favor certain outcomes.

The activity Sampling Rectangles highlights three facts:

1. Subjectively chosen samples may be biased.

2. True random samples are in fact carefully chosen to ensure that every set of a specified size consisting of individuals in the population under consideration is equally likely to be selected.

3. Though an individual event is not predictable, there is an underlying pattern in long-term random behavior.

In the activity, the class collects individual students' results and uses dot plots, as shown in figure 1.2, to display the sample means for each of the two methods used for choosing a sample. Note that the students' collection of the sample means is a "sample" from the distribution of all possible sample means for samples of size 5. The visual display of the two simulated sampling distributions on the overhead transparency allows the class to compare the two methods for selecting a sample.

The population in this activity is the given set of rectangles. Because the entire population is known (which is not usually the case), the population mean—the parameter—of all the rectangles' areas can also be known. It can readily be determined to be 7.42 square units. Draw a vertical line on the blank transparency and line it up through 7.42 on each plot, as shown in figure 1.3, to illustrate how the distributions from the two sampling techniques vary with respect to the actual population mean.

The results of 30 samples (from a class of 30 students) for each of the two methods of choosing samples are given in figures 1.2 and 1.3. In this example, the distribution of the sample mean areas of the samples generated by subjectively choosing "typical" rectangles centers around 11, whereas the distribution of the sample mean areas of the randomly generated samples of rectangles has a much lower center—around 7.5. The subjectively chosen samples have sample mean areas as small as 7 and as large as 16, and the randomly chosen samples have sample mean areas ranging from 3 to almost 14. Box plots such as those shown in figure 1.4 provide another way to compare the two simulated sampling distributions.

The median of the sample mean areas for the subjectively chosen samples of rectangles, like the average of their means, tends to be much higher—around 11.3—than the median of the sample mean areas for the randomly chosen sample of rectangles, which is around 7.3. The population median is 6. The interquartile ranges for both distributions are close to 4, but the locations of the first and third quartiles are shifted to the right for the distribution of the subjective sample means. Notice that the box plots in figure 1.4 show the spread and center but conceal the individual values of the data.

Table 1.1 provides a summary of the statistics for the two sampling methods.

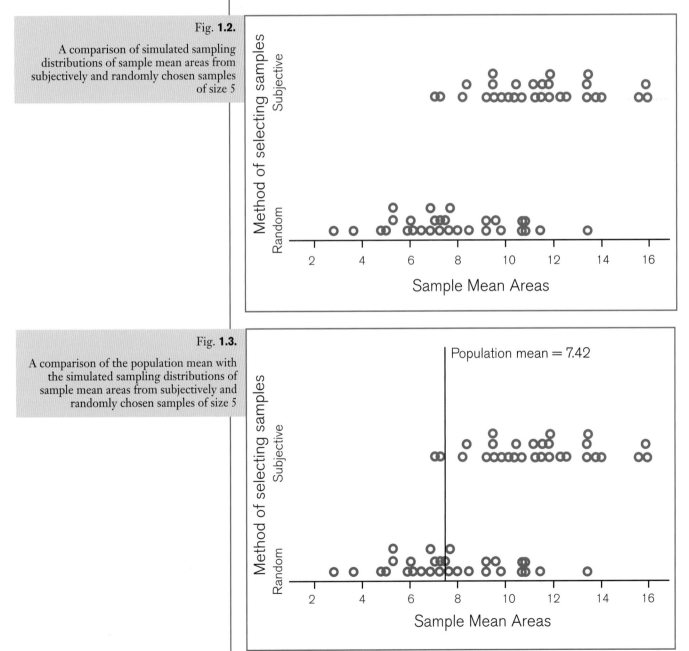

Fig. 1.2.

A comparison of simulated sampling distributions of sample mean areas from subjectively and randomly chosen samples of size 5

Fig. 1.3.

A comparison of the population mean with the simulated sampling distributions of sample mean areas from subjectively and randomly chosen samples of size 5

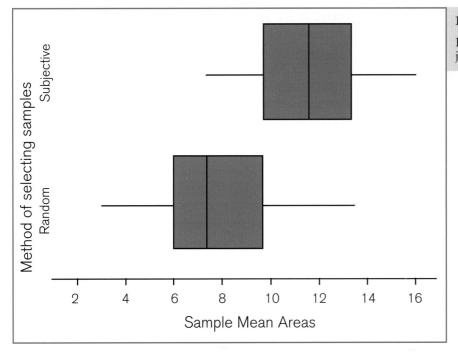

"All students should … be able to display the distribution, describe its shape, and select and calculate summary statistics."
(NCTM 2000, p. 324)

Table 1.1.
Summary Statistics for Sampling Methods

Sample Estimates	Sample Mean Areas of Randomly Chosen Rectangles	Sample Mean Areas of Rectangles Subjectively Chosen as Typical
Mean	7.6	11.3
Median	7.3	11.3
Range	10.6	8.9
Interquartile range	3.6	3.9

Both the dot plots and the box plots reflect the fact that a sample mean area obtained from the randomly chosen rectangles, on average, will be a closer estimate of the population mean area for the 100 rectangles than a sample mean area obtained from the subjectively chosen rectangles. In this example, on average, the subjective sampling method yielded sample means that are higher than the population mean of 7.42. This method is *biased* because it tends to produce samples with estimates for the population means that are higher than 7.42. Estimates are rarely, if ever, equal to the population mean, but a random sample provides an unbiased estimate of the population mean. Sampling strategies that are not based on random selection, such as choosing rectangles by personal judgment of what is typical, have the potential to distort the results by introducing bias.

Variability

Two distributions are of interest in the situation of the 100 rectangles, and each has variability associated with it. The first is the population distribution of the areas of all the rectangles; the second is the sampling distribution of the sample means of the areas of the sampled rectangles from that population. The population distribution of the areas of the set of 100 rectangles has a mean of 7.42. The variability, or

"All students should … be able to understand histograms … [and] parallel box plots … and use them to display data."
(NCTM 2000, p. 324)

The mean absolute deviation, *another measure of spread around the mean, is discussed in Measures of Variability on the accompanying CD-ROM.*

A sample calculation of population standard deviation *appears in Measures of Variability on the CD-ROM.*

"typical" deviation from the mean of 7.42, for the population distribution is measured by the standard deviation.

The *population standard deviation* is calculated by finding the average squared difference between a population value and the population mean and taking the square root to return to the original measurement units of the population values. For the areas of the 100 rectangles, the population standard deviation can be rounded to 5.2.

Sampling distributions also have variability associated with them. Since the example uses a sample of only 30 sample means to *approximate* the sampling distribution of all possible sample means, you cannot calculate the standard deviation of the sampling distribution precisely but must estimate it. To calculate this estimate, you first compute the difference between each observed sample mean and the average of the 30 sample means. Then you square each difference and sum the squares. Because you do not know the mean of the sampling distribution but have only estimated it using the average of the 30 sample means, you divide the sum of the squares by the number of samples (30) minus 1. This subtraction adjusts for the use of a sample mean rather than the population mean itself. Finally, you take the square root and arrive at an estimate of the standard deviation of the sampling distribution.

A relationship exists between the standard deviation of the population distribution and the standard deviation of the sampling distribution of sample means from random samples. Although the focus of this book is not on this relationship, a suggestion about how teachers might use it to make a very important point about randomization is within our scope. Random sampling helps ensure that the variability will be the result of chance and therefore will be predictable. A little stratagem can pique your students' curiosity about this predictability. You can stage the following demonstration.

Before getting your students started on the activity Sampling Rectangles, display a blank sheet of paper and explain that you are going to write something on it that will remain secret until an appropriate time in the activity. In front of the students but without letting them see the actual words, write the following statement:

> For a random sample of size 5 from this population, the standard deviation of the sampling distribution of sample means of randomly selected rectangles' areas can be rounded to 2.3.

The number 2.3, or $5.2 \div \sqrt{5}$, is the standard deviation of the sampling distribution of sample means for random samples of size 5 from the population of 100 rectangles. It is calculated by computing the standard deviation of the population divided by the square root of the sample size. Fold the paper, put it into an envelope, seal it, and give to one of the students for safekeeping.

Once the samples have been collected and discussed, students can use class results to estimate the mean and standard deviation of the sampling distribution of sample means of the randomly generated samples of size 5. With these figures in front of the students, ask the student with the envelope to open it and share its contents with the class. The estimated standard deviation of the sampling distribution of the class's sample means should be close to the number written on the paper. In the case of the data for the randomly selected sample in figure 1.2, the standard deviation of the set of sample means from random samples of size 5 was estimated to be 2.4.

In the activity Sampling Rectangles, students compared simulated sampling distributions of subjectively chosen samples and randomly chosen samples. In Sample Size, the next activity involving the 100 rectangles, students explore the effect of sample size on the sampling distributions.

Sample Size

Goals

- Compare the simulated sampling distributions of means from samples of size 5 and 10 with respect to shape, center, and spread
- Draw conclusions about the effects of sample size on statistical information

Materials and Equipment

- A copy of the activity page for each student
- Copies of the activity sheets "Random Rectangles" and "Data Record" for each student
- The Random Number Generator applet or another random number generator
- Graphing tools that display statistical results

Discussion

In the activity Sample Size, students again consider the 100 rectangles (fig. 1.1), this time examining a larger sample size. Here the students investigate samples of size 10. Although you can suggest other sizes as well, depending on how much time you want to devote to the activity and how you decide to set it up in your classroom, sample sizes should not exceed 10—that is, 10 percent of the population size of 100.

After the students have worked individually to find the sample mean area of five different samples of size 10, they can pool their data and work as a class to plot the simulated sampling distribution of the sample mean areas for the new sample size. (In general, at least 50 samples are needed to get a fairly reliable approximation of the true sampling distribution.) Each group should share results in a way that allows the simulated sampling distributions for the sample means of samples of size 5 and 10 to be compared, by using dot plots or box plots on the same number line. Students can also use graphing software that allows results to be displayed.

The simulated sampling distributions based on 50 samples for each of the sample sizes are illustrated in figures 1.5 and 1.6. Table 1.2 shows the sample statistics for each simulated sampling distribution. The histograms in these figures show that the shapes of the simulated sampling distributions approach a mound or bell shape as the number in the sample increases. In both distributions, the centers are close to the population mean area of all the rectangles—7.42. The box plots in figure 1.7 show that the spread, or variability, decreases as the sample size increases, with the sample interquartile range decreasing from 4.0 to 2.6.

Various questions often arise about sample size. "How many individuals should be sampled?" "What can a sample of 5 or 10 individuals tell about a particular population?" Formal statistical inference explores how confident one can be in reasoning about a population from a sample. Students' work can communicate to them the important message that *as the sample size increases, the variability of the sampling distribution for a sample statistic decreases.*

pp. 94, 98, 99

The Random Number Generator applet on the CD-ROM lets students generate numbers to determine random samples of particular sizes.

"All students should … use simulations to explore the variability of sample statistics from a known population and to construct sampling distributions."
(NCTM 2000, p. 324)

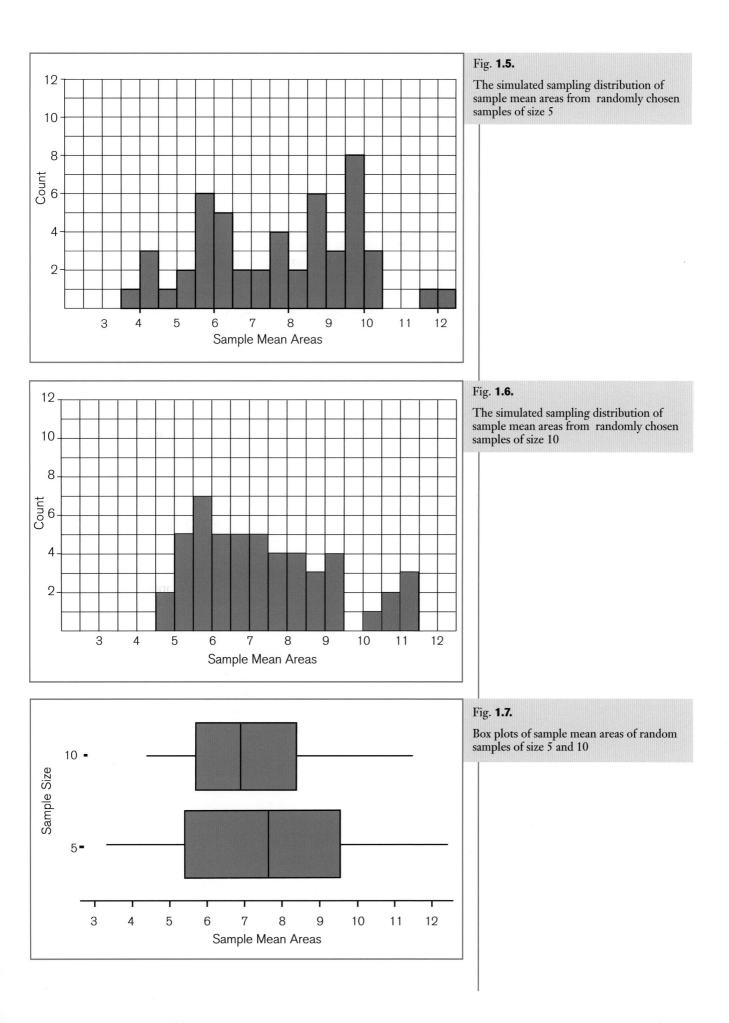

Fig. **1.5.**

The simulated sampling distribution of sample mean areas from randomly chosen samples of size 5

Fig. **1.6.**

The simulated sampling distribution of sample mean areas from randomly chosen samples of size 10

Fig. **1.7.**

Box plots of sample mean areas of random samples of size 5 and 10

Table 1.2.
Statistics for Mean Areas of Samples of Different Size (50 Trials)

Sample Estimates	Sample Size 5	Sample Size 10
Mean	7.78	7.35
Standard deviation	2.38	1.67
Median	7.85	7.05
Interquartile range	4.0	2.6

"All students should ... understand how sample statistics reflect the values of population parameters."
(NCTM 2000, p. 324.

Students can make a numerical comparison between the variability, or spread, of the simulated sampling distribution of the mean areas in their samples of size 5 and the variability of the simulated sampling distribution of the mean areas in their new samples of size 10 by computing the standard deviation or the interquartile range. (Which measure of spread they use may depend on the background of the class.) In the examples, both the estimated interquartile range and the estimated standard deviation decrease as the sample size increases; the simulated sampling distribution becomes "tighter"—that is, in general, the values of the sample mean areas become closer to the population mean area—as the sample size increases. In real situations, statisticians who are deciding on the size of their sample must take into account such factors as resources, time, and cost. The ideal way for them to proceed is to take samples of a size that they have determined to be necessary for the degree of precision that they desire.

The next activity, Sampling Methods, explores two other common sampling methods.

Sampling Methods

Goal

- Use simulations to explore the sampling distributions of stratified random samples and clustered samples

Materials and Equipment

- A copy of the activity pages for each student
- Copies for each student of the activity sheets "Random Rectangles" and "Data Record"
- The Random Number Generator applet or another random number generator
- Graphing tools, including data analysis software

pp. 94, 97, 99, 100–103

Discussion

There are many sampling designs other than simple random sampling. Sampling Methods introduces students to two other common types of samples: *stratified samples* and *clustered samples*. For each set of at least 50 samples, the students use these two new sampling methods to find the sample mean areas of the rectangles. For each simulated sampling distribution, students record a measure of center and a measure of spread (median and interquartile range, or mean and standard deviation) and write a brief description of the shape of the simulated sampling distribution as seen in their histograms.

Have each student gather three or four samples by each method of sampling, and then have the class collect the results. Display them on an overhead transparency (the activity sheet "Distribution of Sample Mean Areas of Rectangles" provides a template) or data analysis software. Again, an alternative is to have the students work in groups, this time assigning different techniques to different groups. Each group would then be responsible for developing its own simulated sampling distribution and sharing it with the class, along with an explanation of what the group did and how the process works. By recording all results on one sheet or transparency, students can inspect the simulated sampling distributions from each method to use in answering the last question on the activity page.

The Random Rectangles applet on the CD-ROM can be used to simulate the sampling process.

Representing Subgroups by Stratified Random Sampling

Simple random sampling, in which every subset of specified size n from the population has an equal chance of being selected, is one method of taking a sample. In some instances, however, researchers want to be sure that certain characteristics are represented in their sample. For example, high school administrators determining how many buses their school would need might want to be sure that all grade levels were represented in any sample of students that they surveyed. It might be the case that juniors and seniors often drove, whereas freshman and sophomores were more dependent on buses. Likewise, a state that wanted to gauge public reaction to a proposed tax change might want to be sure that people from rural areas as well as urban and suburban areas were represented in the sample that it studied.

One sampling technique involves taking what are called stratified random samples. The population is divided into two or more groups, called *strata*, according to some criterion, such as geographic location or grade level, and *subsamples* are randomly selected from each group. This sampling method is most useful when the units in each group are more alike with respect to the characteristics under consideration than are the units in the population as a whole. Because age and income are often good indicators of what a person might purchase, these variables might be used to form strata when conducting a marketing survey of potential customers, for example.

Task A of the activity Sampling Methods illustrates these ideas. Here the "width" of a rectangle is understood as the rectangle's horizontal dimension. Students are asked to suppose that they wanted to be sure that a sample from the population of 100 rectangles had rectangles with both small and large widths. The activity directs them to divide the 100 rectangles into two groups. Table 1.3 lists those with widths less than 3, and table 1.4 lists those with areas greater than or equal to 3.

Table 1.3.

The Stratum of Rectangles with Widths Less than 3, Listed by the Numbers Corresponding to Them

1	2	3	4	5	6	8	9	10	11	12	14	16
18	19	20	21	22	24	27	28	30	32	34	35	36
40	43	44	45	46	48	49	50	51	53	56	62	64
67	68	69	71	73	74	75	77	78	79	80	82	83
84	87	88	89	91	95	100						

Table 1.4.

The Stratum of Rectangles with Widths Greater than or Equal to 3, Listed by the Numbers Corresponding to Them

7	13	15	17	23	25	26	29	31	33	37	38	39
41	42	47	52	54	55	57	58	59	60	61	63	65
66	70	72	76	81	85	86	90	92	93	94	96	97
98	99											

These two strata are not of equal size. The students will find that there are 59 rectangles with widths less than 3 and 41 rectangles with widths greater than or equal to 3. They should randomly choose rectangles from each stratum—that is, 5 small rectangles and 5 large rectangles for a sample of 10. To calculate a sample mean of the rectangles' areas for the combined strata sample, they must use the population proportions, as follows:

$$\frac{59}{100} \bullet \left(\text{mean of a sample from table 1.3}\right)$$
$$+ \frac{41}{100} \bullet \left(\text{mean of a sample from table 1.4}\right).$$

Similarly, using a collection of samples, we can create a simulated sampling distribution and proceed as before to find the sample mean.

Navigating through Data Analysis in Grades 9–12

Cluster Sampling for Scattered Populations

Cluster sampling is often used when studies look across large populations whose members may be widely dispersed and a reliable list of all the individuals or objects in the population is not available. Such studies might include surveys of the level of literacy in the U.S. population or what children in the United States want to do as adults. For studies of this scope, it would be very costly to take a simple random sample of individuals across the United States.

Rather than attempting to collect such a sample, pollsters might randomly select families within a city and then survey every family member. This process produces *clusters* of individuals (those within a family). Rather than testing a random sample of all students in the tenth grade, a high school might randomly select four classes with tenth graders and test all the students in those classes. The classes then would be clusters.

Cluster sampling is useful when the cost of collecting information from the individuals within the cluster is not great. Ideally, each cluster reflects the variability within the population. In reality, individuals within a cluster tend to be more alike than individuals across clusters. Individuals within a household, for example, tend to be more alike than individuals across households. Thus, less information is usually obtained from a cluster sample of size n than from a random sample of size n. However, if the cost of sampling individuals within a cluster is relatively small, it is often possible to collect a larger sample.

In Task B of the activity Sampling Methods, students explore the technique of cluster sampling by considering the 100 rectangles again, this time in clusters that they will sample to make another simulated sampling distribution of the sample mean areas. Figure 1.8 shows a "map" of these clusters, and table 1.5 identifies the rectangles in each cluster.

Table 1.5 gives the area of each rectangle (in parentheses) to show the way in which the areas are distributed in the clusters. The activity sheet shows only the number of the rectangle to reinforce the students' awareness of the need to find each actual rectangle within a cluster and calculate its area.

Table 1.5.
Twenty Clusters of Rectangles*

I	II	III	IV	V	VI	VII	VIII	IX	X
1(1)	3(1)	8(1)	7(12)	19(1)	18(4)	25(18)	40(8)	32(5)	39(12)
2(1)	4(1)	13(14)	11(1)	26(12)	20(1)	34(12)	41(16)	33(12)	47(16)
9(1)	5(1)	14(4)	12(8)	27(4)	21(1)	35(4)	42(6)	38(9)	48(6)
16(1)	6(5)	15(9)	22(4)	28(5)	30(4)	36(4)	43(4)	45(10)	49(10)
17(9)	10(1)	24(5)	23(10)	29(10)	31(16)	37(10)	44(1)	46(3)	50(1)

XI	XII	XIII	XIV	XV	XVI	XVII	XVIII	XIX	XX
51(16)	55(16)	58(4)	68(9)	64(1)	66(10)	79(8)	91(3)	88(4)	84(5)
52(12)	56(4)	59(8)	69(4)	72(12)	75(8)	80(2)	92(4)	89(12)	85(8)
53(6)	57(18)	60(16)	70(18)	73(16)	76(18)	81(15)	93(16)	96(18)	90(16)
54(3)	62(3)	65(5)	71(4)	74(10)	77(3)	86(5)	94(3)	97(4)	99(9)
61(8)	63(9)	67(4)	78(4)	82(6)	83(2)	87(8)	95(6)	98(6)	100(12)

* Rectangles are identified by the numbers corresponding to them in fig. 1.8. The area of each rectangle is shown in parentheses after the number of the rectangle.

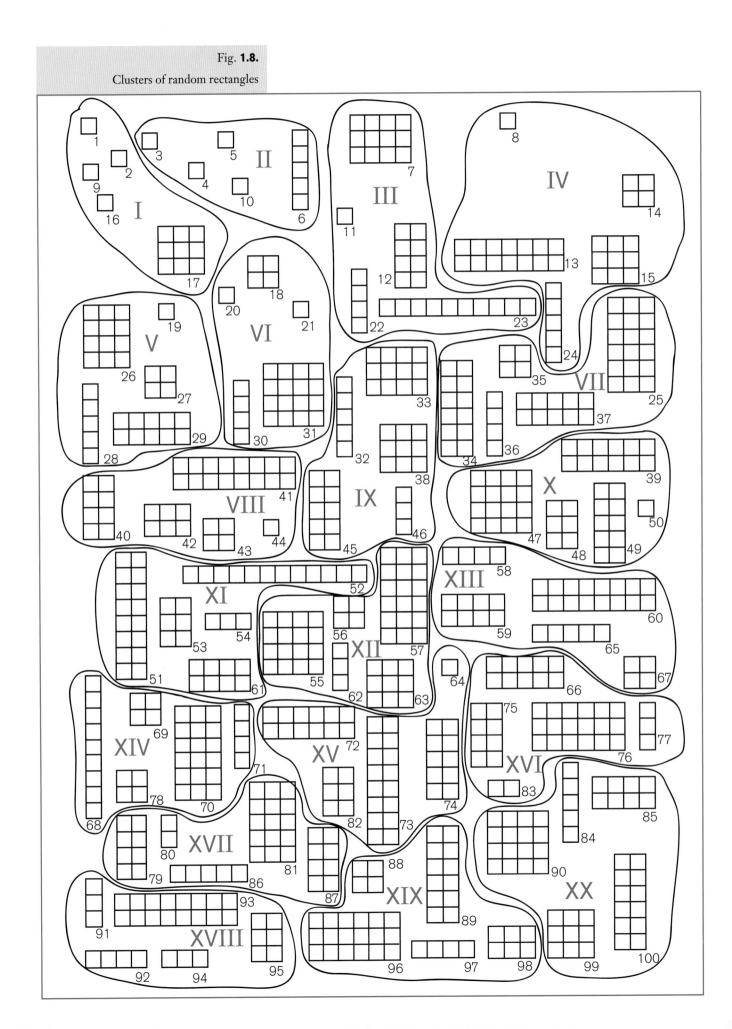

Fig. **1.8.**

Clusters of random rectangles

Figures 1.9 and 1.10 show histograms and box plots for a simulation of 50 samples chosen according to each of the strategies in the activity Sampling Methods. Note in comparing methods that the results change, depending on how effectively strata and clusters have been formed. Students should notice that the sampling method makes a difference in the statistics (see table 1.6).

"All students should … use sampling distributions as the basis for informal inference." (NCTM 2000, p. 324)

Table 1.6
Statistics from 50 Samples Selected by Different Sampling Techniques

Sample Estimates	Simple Random Sample	Stratified Random Sample	Cluster
Sample size	10	10	10
Mean	7.35	7.18	7.35
Standard deviation	1.67	1.35	1.66
Median	7.05	7.26	7.65
Interquartile range	2.30	1.59	1.59

Fig. **1.9.**

The simulated sampling distribution of sample mean areas from samples of size 10 obtained by randomly selecting 5 each from two strata of rectangles

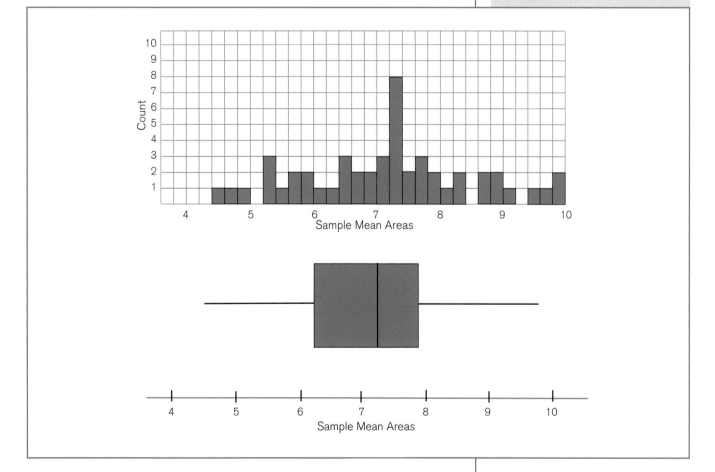

Conclusion

Sampling methods are an important part of making conclusions based on data. Sampling distributions generated by a random process have a predictability that can be used in reasoning about samples and their relation to the population. This regularity allows researchers to understand what is typical and what is not typical in a sampling distribution. Different sampling methods are used for different reasons.

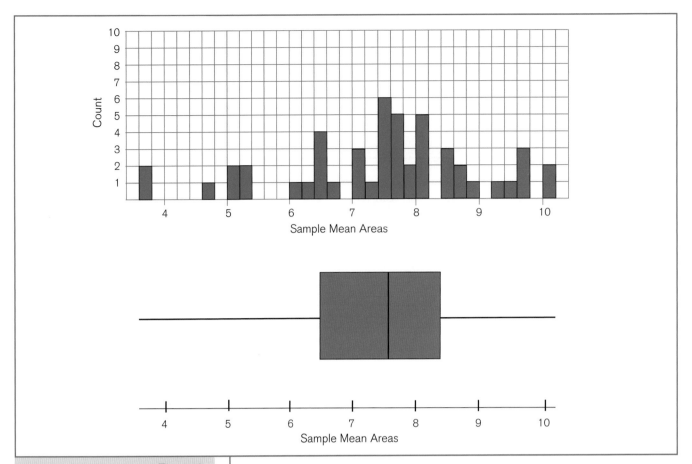

Fig. 1.10.

The simulated sampling distribution of sample mean areas from cluster samples of size 10

Convenience samples are often used in market surveys in shopping malls and on radio call-in shows. Although such samples are typically cheaper to collect than other kinds of samples, they may not reflect the population charactistics being considered. Simple random samples are often costly and difficult to obtain. Stratified random samples, which separate the population into strata, may reduce the cost per observation while also providing estimates for desired strata in the population. Market research firms use stratified random sampling to estimate the potential sales for a new store or to gauge the potential success of advertising campaigns. The U.S. Census Bureau uses cluster sampling to collect national economic and social data. Cluster sampling is an effective technique for gathering specific information at a minimal cost when individual elements within the population are costly to sample, clusters are well defined, and sampling individuals within a cluster is relatively inexpensive once a cluster is obtained.

One of the most important factors in reducing variability—thus ensuring that, on average, an unbiased sample statistic will be relatively close to the population parameter under consideration—is sample size. The previous activities have explored sampling distributions and their behavior. Building on this understanding of sampling distributions and their behavior, the next chapter explores making decisions with categorical data.

NAVIGATING *through* DATA ANALYSIS

Chapter 2
Making Decisions with Categorical Data

In data analysis, a categorical variable represents a characteristic of the population that can be used to classify the individuals or objects into one and only one group (or category). Examples include gender, political party preference, and country of origin.

p. 104

How do you approach a problem or question of interest that has no clear-cut answer? That is, how do you arrive at a conclusion through a decision-making process that involves some degree of uncertainty? The activities in this chapter invite students to explore these questions as they examine an experiment that involves categorical variables. One way to summarize categorical variables in a problem is to count the number of observations in given categories. Analyzing such a problem then usually involves deciding what to do with the counts to determine whether the observation under investigation occurred by chance or resulted from other factors.

This chapter focuses on the problem scenario Discrimination or Not? described at the beginning of chapter 1 (p. 11). As noted in that chapter, statisticians are often asked to look at data when someone believes that discrimination may have taken place. The scenario, which raises a question about possible discrimination, comes from a well-known study conducted by Benson Rosen and Thomas H. Jerdee in 1972 and reported in the *Journal of Applied Psychology* in 1974. Although now thirty years old, the study involves data and statistical thinking that are still relevant today. The activity sheet "Discrimination or Not?" provides the study's scenario, reproduced here for the reader's convenience:

> In 1972, 48 male bank supervisors were each randomly assigned a personnel file and asked to judge whether the person represented in the file should be recommended for promotion to a branch-manager job described as "routine" or whether the person's file should be held and other applicants interviewed. The files were all identical except that half of the supervisors had files labeled "male" while the other half had files labeled "female." Of the 48 files reviewed, 35 were recommended for promotion.

Twenty-one of the 35 recommended files were labeled "male," and 14 were labeled "female." Can we conclude that discrimination against women played a role in the supervisors' recommendations? The three activities that accompany chapter 2 allow students to examine this scenario. As they work, they will identify relevant questions to consider about discrimination, create a simulation to investigate the possible presence of discrimination, define the numerical summary or statistic that they will calculate from each simulation, and form and explore the simulated sampling distribution based on the statistics resulting from each simulated sample.

To help your students begin thinking about the scenario, you might give them copies of the activity sheet "Discrimination or Not?" and ask them to brainstorm about the following question: "How could you create a model to determine whether any given distribution of the 35 "recommended" files into "male" and "female" files could have occurred by chance variation?" Students might work in groups to formulate their strategies and then reconvene as a class to discuss how they would use these ideas to analyze the problem. After considering with your students some possible models to evaluate any results from the study, you could turn their attention to the first activity, What Would You Expect?

What Would You Expect?

Goals

Begin a statistical exploration of Discrimination or Not? by completing two-way tables showing hypothetical situations with—

(*a*) discrimination clearly playing no role in the results

(*b*) discrimination against women clearly playing a role in the results

(*c*) the results falling in a "gray" area

Materials

- A copy of the activity pages for each student
- A copy of the activity sheet "Discrimination or Not?" for each student

Discussion

Understanding the problem is essential to solving it. Our scenario involves two categorical variables of interest: *gender* (male, female) and *recommendation* (recommended, not recommended). Gender, the *explanatory variable*, might help explain or provide information about recommendation, the *response variable*. The activity helps students use two-by-two (or two-way) tables such as tables 2.1 and 2.2 to organize the information from the study and to consider different hypothetical scenarios. If the recommendations involved no discrimination, one could reasonably expect that the number of males recommended for promotion and the number of females recommended for promotion would be about the same because the numbers of files for each category of gender were the same. (See table 2.1.) Note that either 17 or 18 males recommended for promotion could be regarded as approximately half of the 35 individuals who were recommended for promotion. The number 17 has been used in the table here.

Table 2.1.
A Hypothetical Situation Involving No Evidence of Discrimination against Females

	Recommended for Promotion	Not Recommended for Promotion	Total
Male	17	7	24
Female	18	6	24
Total	35	13	48

For the table constructed in the activity, gender defines the rows, and recommendation defines the columns. Within the groups of males and females, it is important to consider the percentage in each cell.

By contrast, if the recommendations showed strong evidence of discrimination against the women considered for promotion, the numbers in such a table would look quite different. The activity asks students to complete a table showing the situation if the vast majority of the males were recommended for promotion. Table 2.2 shows the case with *all* the males recommended for promotion. Students may also suggest 22 and 23 as reasonable values for the number of men recommended for promotion if discrimination against women clearly played a role.

pp. 104, 105–6

"All students should … formulate questions that can be addressed with data and collect, organize, and display relevant data to answer them."
(NCTM 2000, p. 324)

Table 2.2.
A Hypothetical Situation Involving Strong Evidence of Discrimination against Females

	Recommended for Promotion	Not Recommended for Promotion	Total
Male	24	0	24
Female	11	13	24
Total	35	13	48

The activity then calls on students to suppose a more complicated hypothetical situation, with the evidence of discrimination against the women falling into a "gray" area. In such a case, the presence of discrimination would not be obvious without further investigation. Students may offer a variety of tables for this case, depending on how they perceive the possible variation around the expected values. Table 2.3 illustrates such a situation with the data from the original study.

Table 2.3.
Actual Results of the Discrimination Study

	Recommended for Promotion	Not Recommended for Promotion	Total
Male	21	3	24
Female	14	10	24
Total	35	13	48

As the students consider the three scenarios, they should reflect on one of the main questions of interest: How much variability would one expect in the observed counts in a table like table 2.1, which shows the situation when no evidence of discrimination exists? Is it likely that 19 men would be recommended for promotion when you expected 17 or 18? Is it likely that 21 men would be recommended? If this study were repeated many times and discrimination played no part in the recommendations, what counts would be observed? How would the numbers of males and females recommended for promotion fluctuate just by chance variation in the presence of no discrimination?

Simulating the Discrimination Case

In real life, researchers do not usually have the opportunity to repeat a study many times. Simulation, however, is a powerful tool that allows them to model situations and investigate what would happen if they could repeat the study many times. The next activity, Simulating the Case, helps students use simulation to investigate the scenario Discrimination or Not? By allowing students to gather information about the variability that can result by chance when there is no discrimination, the simulation prepares them to use statistical reasoning to answer the real question about discrimination.

Simulating the Case

Goals

- Investigate Discrimination or Not? by—
  (*a*) creating a simulation to model a review of 48 personnel folders
  (*b*) repeating the simulation 20 times to get an idea of the variability to expect from chance when there is no discrimination
- Use the results of the simulation to consider whether the number of males recommended for promotion was the result of chance variation or the result of discrimination against women

Materials

- A copy of the activity pages for each student
- A copy of the activity sheet "Discrimination or Not?" for each student
- A standard deck of 52 playing cards for each student or pair of students

pp. 104, 107–8

Discussion

In Simulating the Case, students use a standard deck of 52 playing cards to create a simple simulation of the situation in "Discrimination or Not?" They remove 2 red cards and 2 black cards, leaving 48 cards divided evenly between black and red. The students can then let the 24 black cards represent the male candidates for promotion and the 24 red cards represent the female candidates, or vice versa.

The structure of the simulation depends on keeping the row and column totals from tables 2.1, 2.2, and 2.3 fixed: That is, (1) after a review of 48 candidates' files, 35 out of the 48 candidates were recommended for promotion, and (2) 24 of the files were labeled "male" and the other 24 were labeled "female."

The object of the activity, then, is to use the 48 cards to simulate the selection process that resulted in the choosing of 35 candidates to recommend for promotion from the 48 personnel files in the study if no discrimination were present. An alternate method involving quick and easy counting is to investigate the 13 candidates who were not recommended for promotion. Dealing 13 cards is simpler and more efficient than dealing 35. Students using this method can then quickly count the cards that represent men who were not recommended for promotion and subtract this number from the total number of male candidates reviewed to get the number of men recommended for promotion. For example, if in a simulation 8 out of the 13 who were not recommended for promotion were men, then the other 16 men would all have been recommended for promotion. In this simulated sample, 16 is the numerical summary, or statistic, that estimates the number of men that one would expect to be recommended for promotion if there were no discrimination.

Simulated counts generated by such a chance process will produce a simulated sampling distribution of counts for males recommended for promotion if there were no discrimination. Students should perform

Recall that a *sampling distribution* is the distribution of the values of a statistic for repeated samples of the same size from a population. For the scenario Discrimination or Not? the number of black cards (representing the number of males recommended for promotion) from each simulation is the *statistic*.

the simulation at least 20 times to get a sense of the center, shape, and spread of this simulated sampling distribution. The observed number of males recommended for promotion (21) can then be compared to the simulated sampling distribution.

You can ask your students to carry out the simulation in one of several ways. As a homework activity, students can independently create their own sets of 20 trials of the simulation—that is, they can deal out samples of 13 cards 20 times, being sure to shuffle the deck thoroughly between deals. Then each student will have a simulation of the sampling distribution of sample counts of males recommended for promotion to analyze. Alternatively, students might conduct only one or two trials independently before the whole class works together to pool results and analyze them. Another approach is to assign the activity for students to complete in pairs.

Direct your students' attention again to the tables in the activity What Would You Expect? and point out that only one of the observed counts in the table cells can vary freely, because the row and column totals are fixed. For example, once the number of males recommended for promotion has been determined, the other three cell values must be counts that add to the fixed row and column totals. Thus, data analysis can be based on just one of the cell categories. Here the number of males recommended for promotion is the category used.

Chapter 1 introduced ideas about sampling, sampling distributions, and variability and explained the concept of a *statistic*. The statistic being considered here is the number of men recommended for promotion. The observed value of this statistic from the actual study is 21. In the activity Simulating the Case, students use simulation to create a simulated sampling distribution of the statistics obtained from each simulation. The overriding goal of the activity is to provide information to answer the following question: "Is the *observed* number of males recommended for promotion greater than the number that would be *expected* as a result of chance variation?"

The next activity, Analyzing Simulation Results, presents three sample student simulations that teachers can use as an exercise for students to work on individually or together as a whole class.

Analyzing Simulation Results

Goals

- Continue to investigate the scenario Discrimination or Not? by comparing and analyzing the shape, center, and spread of the simulated sampling distributions based on three sets of student simulations
- Use the simulated sampling distributions to consider how infrequent an event must be for it to be regarded as "rare"

Materials

- A copy of the activity pages for each student
- A copy of the activity sheet "Discrimination or Not?" for each student

Discussion

Figures 2.1, 2.2, and 2.3 present the results of three sets of student simulations that can help answer the question about discrimination raised by the scenario Discrimination or Not?

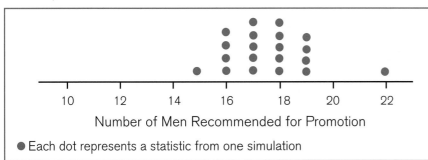

Number of Men Recommended for Promotion

● Each dot represents a statistic from one simulation

Descriptions of any distribution should focus on the shape, center, and spread or variability of the distribution. Students can see that the shape of the simulated sampling distribution in figure 2.1 is approximately symmetrical, with one unusual observation at 22 males. The value at 22 is a potential *outlier*—that is, an observation that does not fall into the overall pattern of the other observations in the distribution. Half of the counts are higher than the median of 17.5 males. The mean, or average, of the simulated sampling distribution appears to be around

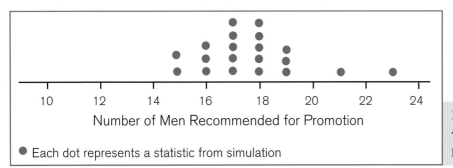

Number of Men Recommended for Promotion

● Each dot represents a statistic from simulation

Statisticians would formalize the question of interest in Discrimination or Not? in two statements: *a null hypothesis,* which would assume that discrimination played no part in the recommendations and that any departure from table 2.1 was due solely to chance variation, and an *alternative hypothesis,* which would assume that discrimination against women played a role in the outcome.

pp. 104, 109–12

"All students should ... select and use appropriate statistical methods to analyze data for univariate measurement data, be able to display the distribution, describe its shape, and select and calculate summary statistics."
(NCTM 2000, p. 324)

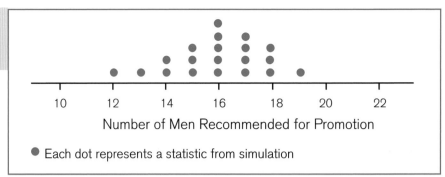

Number of Men Recommended for Promotion

● Each dot represents a statistic from simulation

Recall that the *interquartile range* is the distance covered by the middle 50 percent of the values of a data set. To calculate the interquartile range (IQR), first order the data from lowest to highest values. Then divide the set into two equal groups at the "middle." If the set contains an even number of values, divide it at the median. If the set contains an odd number of values, make your groups from the points on either side of the median. Next, find the medians of the low and high "halves." These medians are the first and third quartiles (25th and 75th percentiles), often denoted as Q_1 and Q_3, respectively.

$$IQR = Q_3 - Q_1$$

The sample *standard deviation* (SD) of a data set is determined by—

- finding the difference between each observation and the sample mean,
- squaring the differences to eliminate negative values,
- summing the squares of the differences,
- dividing the total by the number of values in the set *minus one*, and
- taking the square root of the quotient.

We can express this calculation as

$$\sqrt{\frac{\left(x_1 - \bar{x}\right)^2 + \left(x_2 - \bar{x}\right)^2 + \ldots + \left(x_n - \bar{x}\right)^2}{n-1}},$$

where the set has n data values, x_1, x_2, ..., x_n, and $\bar{x} =$ the sample mean.

18 males and in fact can be calculated as 17.6 males. The counts vary from 15 to 22; however, most of the observations fall between the values of 16 and 19. The variability in the observations can be described by the interquartile range (IQR), or 2; the sample standard deviation (SD), or 1.6, or the sample mean absolute deviation (MAD), or 1.2 males.

The simulated sampling distribution in figure 2.1 conforms to what we might expect to occur in the selection process if there were no discrimination. If the process of selecting applicants for promotion occurred as modeled by the cards, then we would expect about half the promotions to be male and the other half to be female. Therefore, out of 35 cards, the expected number of black cards would be 17.5. The mean of the 20 sample counts was 17.6. The simulated sample distribution is approximately symmetrical to a vertical line through the center of the distribution.

The results of the simulations in figure 2.1 do seem to provide evidence that the selection of 21 males for promotion was not due to chance variation with no discrimination. Only 1 result of the 20 simulated counts was 21 or higher. Thus, for these data, an estimate of the probability of obtaining 21 or more black cards would be 1/20, or 5 percent. The selection process modeled by the cards was chance variation with no discrimination. Is 5 percent small enough to support a claim of discrimination? Do the simulated results in figure 2.1 supply enough evidence to support the possibility that selecting 21 males out of 35 recommended for promotion was not due merely to chance but instead was due to discrimination against women?

Figure 2.2 shows a second set of 20 simulated counts. On the basis of these 20 simulated results, the estimated chance of obtaining 21 or more black cards would be 2/20, or 10 percent. The card selection process represents chance variation with no discrimination. Is 10 percent small enough to support a claim of discrimination against females? How small is "small enough"? What intuitively is "small enough"? Most people are comfortable about saying that an event that occurs no more often than 1 out of 20 times, or 5 percent of the time, is a rare event. Most people would be less confident about calling an event that occurs as often as 2 out of 20 times, or 10 percent of the time, rare.

How small is "small enough"?

You can use two simple exercises to help give your students a sense of "how small is small enough"—that is, how small the probability of an event occurring needs to be for the event to be considered rare. The first exercise calls for two new decks of cards in sealed boxes. Before class, you should—

- open the boxes carefully (so that they can be resealed) and remove the cards, separating red and black cards;
- put all the red cards (including jokers) in one box and all the black cards (including jokers) in the other box;
- reseal the boxes, noting which box has the black cards and which one has the red cards;
- take the box containing the red cards to class and offer a prize for anyone who draws a black card;
- break the seal in front of the class and remove the jokers (reinforcing the illusion of a sealed, intact, new deck of cards);
- direct the students to line up to draw a card. (To suggest that you are keeping the probability of drawing a black card constant for each draw, replace the card that each student draws and reshuffle the deck.).

It usually takes only about five students drawing red cards one after the other for a class to suggest that the deck does not contain any black cards. Since a student is drawing a card from a 52-card deck, the probability of drawing a black card should be 26/52, or 1/2, for each draw. The probability of drawing 5 red cards in a row is $(1/2)^5$, or 3.125 percent. Students will suspect that observing 5 red cards in a row is not likely to happen by chance alone; the probability that you would observe 5 red cards in a row is less than 5 percent.

The second exercise uses a two-headed coin to help students develop a sense of how infrequent an event has to be to be considered rare. Two-headed coins are available in kits of magic tricks or in the shops that sell them. For this exercise—

- bring such a coin to class and hold it up for the students to see, being careful not to reveal that both sides are heads;
- call heads, and then toss the coin, asking for a student to call tails;
- flip the coin in the same manner several times, reporting the results each time;
- Keep flipping the coin until the students conclude that heads appear on both sides.

Usually, as with the cards, this process takes only about five turns. If a coin is fair (head on one side, tail on the other, with each side having a 50 percent chance of landing face up after a toss), the probability of observing 5 heads in a row would be $(1/2)^5$ or 3.125 percent. Once again, most students will intuitively suspect that something rare is happening when five tosses produce 5 heads in a row. The probability of this happening by chance is again less than 5 percent, which in fact is a typical benchmark that statisticians use for calling an event rare. That is, in statistics, an event is often considered rare if it happens by chance 5 percent of the time or less. (In some situations, 5 percent may not be appropriate; the situation may help dictate the percentage.)

Figure 2.3 displays a third set of the counts of the number of men recommended for promotion from 20 simulations. The simulated sampling distribution here is approximately symmetrical, with a slight tail (*skewness*) to the left. Most of the counts are from 15 to 18. There are no counts higher than 19. As estimated in this chance process, if no

The sample *mean absolute deviation* is the average distance of the values in a data set from the mean. Expressed mathematically, the mean absolute deviation of a data set is equal to

$$\frac{|x_1 - \bar{x}| + |x_2 - \bar{x}| + \ldots + |x_n - \bar{x}|}{n},$$

where the set has n values, x_1, x_2, ... x_n, and x is the sample mean.

The probability of observing 21 or more black cards if the selection process is due to chance variation is called the *p*-value. Small *p*-values provide strong support for the alternative hypothesis—in our situation, the hypothesis that discrimination against women could have played a role in the selection process.

 "All students should … develop and evaluate inferences and predictions that are based on data by using simulations to explore the variability of sample statistics from a known population."
(NCTM 2000, p. 324)

discrimination were occurring, the probability of observing 21 or more males would be 0. This simulated sampling distribution provides strong evidence that the selection of 21 or more males recommended for promotion was not due to chance. Note that the mean and median of the simulated sampling distribution are approximately 16, lower than expected and different from the centers in figures 2.1 and 2.2. These results could have occurred as a result of the random nature of dealing the cards, but a possibility also exists that the student carried out the simulation improperly.

Instances of 21, 22, or 23 females could happen by chance variation with no discrimination against women. Having a disproportionate number of females is not out of the ordinary for this chance process. What is unusual here is where the distribution is centered.

How stable are the results?

Figures 2.1, 2.2, and 2.3 illustrate the fact that the simulated sampling distributions of 20 simulated counts of the number of men recommended for promotion will fluctuate from one simulated sampling distribution to another. Combining all the student simulations into one simulated sampling distribution will stabilize the sampling distribution of counts and help students understand what can be expected to happen in the long run. Figure 2.4 illustrates a data set of 1291 student simulations.

This larger set of simulated counts provides a relatively stable model of the shape of the sampling distribution of the number of black cards selected out of 35 to represent the number of men recommended for promotion. The shape is approximately symmetrical, centering at a median of 17 and a mean of approximately 17.5. The variability is similar to that observed in the previous simulated sampling distributions, with an interquartile range of 3 and a standard deviation of about 1.6.

Equipped with the information from the 1291 simulations, you and your students can now reconsider the original question: Is there evidence of possible discrimination against females? Does there appear to be evidence to support a claim that the recommendation of 21 males out of 35 candidates for promotion (black cards) was not due to chance variation? Figure 2.4 displays 36 out of 1291 counts that were 21 or

Fig. **2.4.**

The number of men recommended for promotion in 1291 simulations

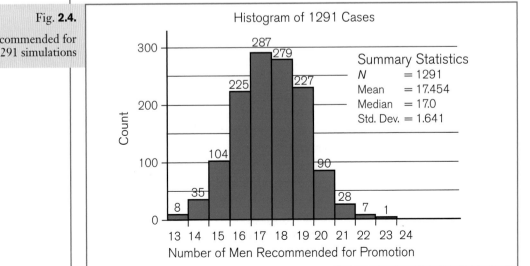

Histogram of 1291 Cases

Summary Statistics
N = 1291
Mean = 17.454
Median = 17.0
Std. Dev. = 1.641

Number of Men Recommended for Promotion

higher. Thus, if no discrimination were occurring, the simulated probability of observing 21 or more black cards (that is, the chance of recommending 21 or more males for promotion) would be 36/1291, or 2.79 percent, assuming that dealing the 35 cards was a random process.

This result would seem to provide evidence to support a claim that recommending 21 men out of 35 candidates was not due to chance variation but could have been due instead to discrimination against women. However, it is worth noting that if the number of men recommended for promotion in the study had been 20—just one fewer than the actual number of 21—we would be trying to answer the question of interest by looking at the simulated probability of observing 20 or more black cards, or 126/1291, which is 9.52 percent. This percentage is obviously substantially higher than the selected benchmark of 5 percent for a rare event, and in such a case, we would be less sure about advancing a claim of discrimination against women.

How reliable are the results?

How reliable or consistent are the simulated class results from one class of students to another? How do they compare to previous results? Breaking down the class data set in figure 2.4 into the two school terms from which the results were obtained illustrates the reliability of the simulated results. Figure 2.5 displays the simulated sampling distributions of the results for the two different groups of students who participated in this activity.

Even though figure 2.5a has fewer simulated results than figure 2.5b (278 compared to 1013), the shapes of the two simulated sampling distributions are very similar, and both are approximately symmetrical. They are both centered at a mean of approximately 17.5. Measures of the variability of the two simulated sampling distributions are also similar, with interquartile ranges of 3 (from 16 to 19) for both and standard deviations of 1.62 and 1.64. In figure 2.5a, the chances of obtaining 21 or more black cards were 8 out of 278, or 2.9 percent. In figure 2.5b, the chances were 28 out of 1013, or 2.76 percent. The simulated results appear to be reliable and consistent—that is, the results are repeatable.

Conclusion

This chapter's examination of the scenario Discrimination or Not? focused on making a decision about discrimination in a situation where there was variability and uncertainty. The question was whether the outcome observed in the study—21 men recommended for promotion out of 35 recommended candidates—could reasonably be attributed to chance variation or was more likely to be due to some other effect, such as discrimination against female candidates. By exploring the data graphically and using numerical summaries and simulated probabilities, you and your students will have discovered that you can deal statistically with the variability (uncertainty) of the data and make a statistical decision about the case.

Two possible errors can enter into the decision-making process of statistical inferences. Remember that the reasoning behind any claim of discrimination against females in the bank supervisors' recommendations rests on the fact that the likelihood of actually recommending 21 or

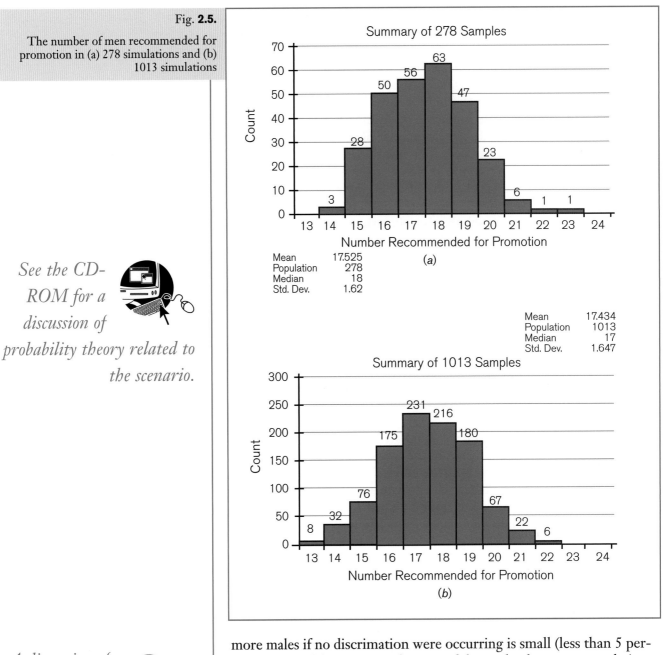

Mean 17.525
Population 278
Median 18
Std. Dev. 1.62

(a)

Mean 17.434
Population 1013
Median 17
Std. Dev. 1.647

(b)

See the CD-ROM for a discussion of probability theory related to the scenario.

A discussion of both types of errors—type I and type II—is included on the CD-ROM.

more males if no discrimation were occurring is small (less than 5 percent). What if, however, in the case of the study, the recommendations of the 21 male candidates happened to be part of the approximately 5 percent of such results that occur by chance? On the one hand, we could potentially make a mistake in claiming that discrimination against women played a role in the bank supervisors' decisions when it did not. This would be an example of a *type I error*. Errors of this type occur when the null hypothesis—in our case, that there was no discrimination—is rejected but actually is true.

On the other hand, we could potentially make a mistake of a different type—*a type II error*—by not supporting a claim of discrimination against women when such discrimination existed. This type of error occurs when the null hypothesis—in our situation, the hypothesis that there was no discrimination—*should* be rejected in favor of the alternative hypothesis—here, that there was discrimination against women—but is not.

At the beginning of chapter 2, the activities introducing the scenario Discrimination or Not? raised questions about the design of the study. What important assumptions were made, and how did they affect the outcome? What elements were necessary to make this a well-designed study? One assumption was that the recommendation of each bank supervisor was an individual decision and that supervisors did not discuss the files with one another or think that other supervisors would be looking at the files. For statistical purposes, there was a need to assume *independence* among the bank supervisors (which actually existed in the original study).

Other questions are also relevant. How were the files assigned to particular supervisors? Would it have been important to use a random process in deciding whether a supervisor got a file labeled "male" or a file labeled "female"? Why, or why not? How might statistical decisions about possible discrimination have been affected if more than 48 supervisors had participated in the study?

In the actual study, except for the genders assigned to the candidates, all the files contained the same information. In a real-life situation, it is likely that there would be 48 completely different files under consideration. It would not be reasonable to assume that the files represented equally qualified candidates. A statistical analysis would have to find a way to incorporate the differences among the applicants. If an actual discrimination suit against a bank were being tried in a court of law, a jury would need to consider, among other things, whether the bank had a history of discriminatory practices.

The questions and statistical issues raised in this chapter will be explored further in the following chapters. It is important to keep in mind that in all the activities involving the scenario Discrimination or Not? statistical analysis does not prove the presence or absence of discrimination against women; it only sheds light on expected behavior under chance variation when no discrimination has occurred.

NAVIGATIONS SERIES

GRADES 9–12

NAVIGATING *through* DATA ANALYSIS

Chapter 3
Making Decisions with Numerical Data

For a discussion of different types of graphs, see Navigating through Data Analysis in Grades 6–8 *(Bright et al. 2003, pp. 14–16).*

"All students should … understand the meaning of measurement data and categorical data, of univariate and bivariate data, and of the term variable."
(NCTM 2000, 324)

Data analysis involves obtaining information in the presence of variability. Chapter 2 examined a problem in which the variables were categorical (the classifications "recommended for promotion," "not recommended for promotion"; male, female). This chapter looks at numerical data, which are quantitative and are also known as measurement data. Visual tools that may prove useful in analyzing this type of data include histograms, scatterplots, side-by-side box plots, and dot plots.

Displaying data appropriately is only part of the job of data analysis. More important are the abilities to describe the data plots and interpret those descriptions in context. Arriving at meaningful interpretations requires knowing what question or questions one seeks to answer with data. Thus, perhaps the most important task is deciding exactly what it is that one wants to know. To do this, it is essential to ask good questions. It is best to formulate key questions before data are collected, but statisticians are often put in the situation of studying data after the fact.

Like chapter 2, this chapter centers on a single problem scenario—this time from a study of diet and cholesterol. The activities focus on (1) analyzing the data from the scenario Dietary Change and Cholesterol in light of a question of interest and (2) moving from representing the data graphically to drawing a conclusion about the results, much as the activities in chapter 2 did with data from Discrimination or Not? There is no random sampling or random assignment of treatments, so this is a purely observational study. The scenario Dietary Change and Cholesterol and the resulting data are reproduced here from the blackline master:

p. 113

High cholesterol is a contributor to heart disease. Table 1 [shown here as table 3.1] lists data from a study investigating the effect of dietary change on cholesterol levels. Twenty-four hospital employees voluntarily switched from "a standard American diet" to a vegetarian diet for one month. The data show their cholesterol levels both before and after the dietary change, in milligrams of cholesterol per deciliter of blood (mg/dL). Suppose for the activities that it is always desirable to decrease the level of cholesterol in the blood. Thus, assume that the purpose of the switch to the new vegetarian diet is to decrease that level.

Table 3.1.
Cholesterol Levels before and after Changing Diets

Before (mg/dL)	After (mg/dL)	Before (mg/dL)	After (mg/dL)
195	146	169	182
145	155	158	127
205	178	151	149
159	146	197	178
244	208	180	161
166	147	222	187
250	202	168	176
236	215	168	145
192	184	167	154
224	208	161	153
238	206	178	137
197	169	137	125

(Rosner 1986)

pp. 113, 116

The first activity, What Can You Know—How Can You Show? emphasizes the importance of asking good questions and organizing and displaying data appropriately to answer the questions.

If your students are already well acquainted with graphical displays, you may want to substitute the alternate activity Data-Based Dietary Decisions for the activity What Can You Know—How Can You Show? The two activities share goals and materials, but Data-Based Dietary Decisions allows students to formulate and carry out their own solutions with little additional direction. After giving students time to work on this activity, you can use the commentary provided in the discussion of What Can You Know—How Can You Show? to conduct a class critique of the students' work. Alternatives to having students develop solutions of their own include having students critique one or more of the sample solutions provided or giving more explicit direction about the particular kind of analysis that they should use.

What Can You Know–
How Can You Show?

Goals

- Examine a data analysis problem from different perspectives
- Formulate appropriate questions that can be answered by data analysis
- Use appropriate graphical displays to make informed decisions in the presence of variability
- Compare different analyses of a single scenario

Materials and Equipment

- A copy of the activity pages for each student
- A copy of the activity sheet "Dietary Change and Cholesterol" for each student
- A statistical graphing utility or graph paper

pp. 113, 114–15

Discussion

Researchers' formulation of the problem—the question that they want to answer with data—is crucially important in determining the analysis that they will undertake and the plots that they will use to display the data. What question (or questions) does the study described in the scenario Dietary Change and Cholesterol have the potential to answer? Consider the following: "Is the diet effective?" "For what patients, if any, will the diet make a difference?" "Can a doctor predict, with some degree of certainty, the change in a patient's cholesterol after following the diet?" "Could you as a patient predict your own change in cholesterol after being on the diet?"

The data in the study can be used to answer such questions, and graphical displays can be important tools in discovering and presenting the solutions. The activity What Can You Know—How Can You Show? explicitly focuses on the relation between the question of interest and a graph that may help answer that question. Students should make direct connections between the questions that they pose and the plans that they make for using and displaying data, matching "what to look for" with what they claim to want to know.

A major goal of this activity is to develop students' abilities to ask precise questions and to plan analyses that answer them. As students work through the activity, they should specify the features of the graphs that they use, indicating how those features help answer the questions. Finally, on the basis of their analyses, they should be able to answer for themselves the question of whether they would change from a standard American diet to a vegetarian diet to lower their cholesterol levels. They should also be able to say whether, if they were either doctors or dietitians, they would recommend that their patients change from a standard American diet to a vegetarian diet to lower their cholesterol levels.

"All students should … formulate questions that can be addressed with data and collect, organize, and display relevant data to answer them."

(NCTM 2000, p. 324)

As students begin their analyses, they may need to refine their questions by quantifying them. Considering such phrases as "how much?" or "how many?" or "how often?" can help them do this. A series of typical questions and answers based on different graphical representations is provided here. It includes questions that someone might ask about the results, and the questions are accompanied by discussions of graphical displays that could be used to answer them. The discussions highlight strengths and weaknesses of each of the displays as an answer to a particular question. The strategies and graphs are *not* recommended solutions. Rather, along with the associated commentaries, they are intended to guide you in leading discussions about students' work. In particular, as noted in the discussions, some of the graphs illustrate limited approaches to the problem. The chapter ends with two ways to simulate sampling distributions to investigate whether the results observed in the study could be due to chance or are unusual enough to suggest that other factors, such as a vegetarian diet, might be involved.

Does the diet make a difference?

Both health-care providers and patients would be interested in knowing whether the diet made a difference in the study. Using a scatterplot with "before" and "after" coordinates (*Before, After*) such as that shown in figure 3.1 is one approach to the question.

Fig. **3.1.**

A scatterplot of the data (*Before, After*) from table 3.1

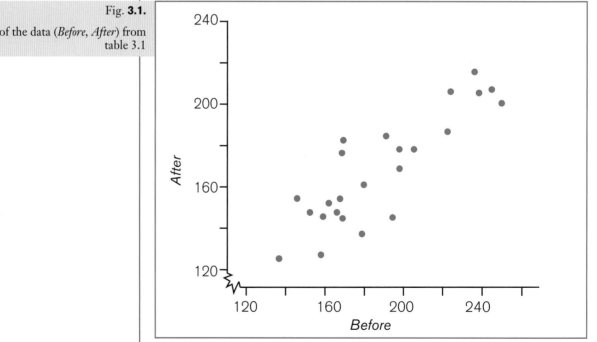

If the diet clearly made no difference, the cholesterol levels would not change with a new diet, and the ordered pairs would reflect no change. In other words, the points would all be on the line *After = Before* (the line *y = x*). Plotting this line can provide a benchmark for analyzing changes in cholesterol as a result of diet, as illustrated in figure 3.2. If there were no changes at all, the data would look like that in figure 3.3.

If the diet clearly had an effect and the cholesterol level of everyone in the study improved (that is, decreased), the plot could look like that

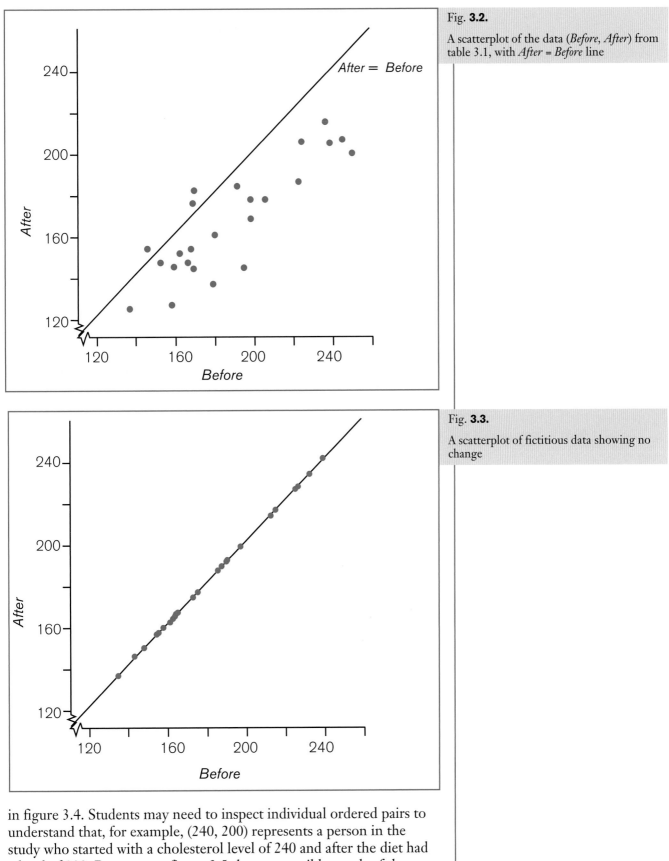

Fig. **3.3.**

A scatterplot of fictitious data showing no change

in figure 3.4. Students may need to inspect individual ordered pairs to understand that, for example, (240, 200) represents a person in the study who started with a cholesterol level of 240 and after the diet had a level of 200. By contrast, figure 3.5 shows a possible graph of the opposite scenario—a diet that resulted in increased cholesterol levels for everyone.

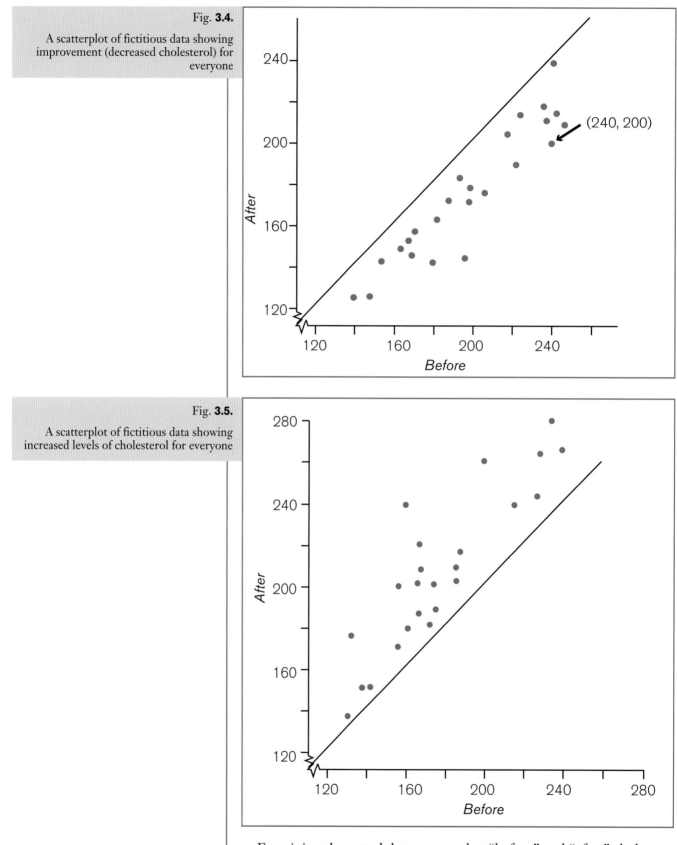

(240, 200)

Examining the actual data, we see that "before" and "after" cholesterol levels show a decrease for most of the volunteers, although other questions, such as to what degree the levels improved, remain unanswered. One approach to the question "Does the diet make a

Navigating through Data Analysis in Grades 9–12

difference?" is to treat the "before" and "after" data as two independent samples. In this approach, side-by-side box plots, dot plots, or aligned histograms can be used for comparisons. In the best of all worlds, box plots answering the question "Does the diet make a difference?" would look like those in figure 3.6, which clearly show improvement for all the participants in the study.

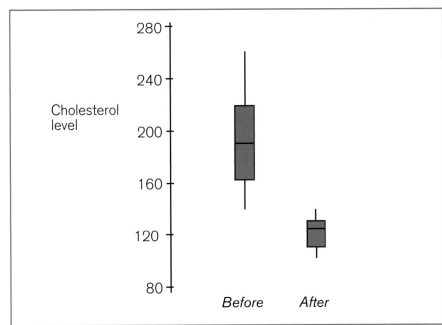

Fig. **3.6.**

Box plots showing improvement

Figures 3.7 and 3.8 show the data from the study in side-by-side box plots and dot plots, respectively. Neither figure shows results that are clear-cut. In both, the center of the distribution of "after" levels is below that of the "before" values. The sample mean dropped by about 19 mg/dL, and the sample median decreased by 14 mg/dL. However, there is a great deal of overlap and variability among the data in both sets. The two variables have sample ranges of 113 mg/dL and 90 mg/dL, with *Before* extending from 137 mg/dL to 250 mg/dL, and *After* extending from 125 mg/dL to 215 mg/dL. The overlap makes any improvements in level difficult to determine from these analyses.

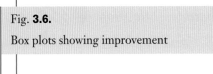

All students should … understand histograms, [side-by-side] box plots, and scatterplots, and use them to display data."
(NCTM 2000, p. 324)

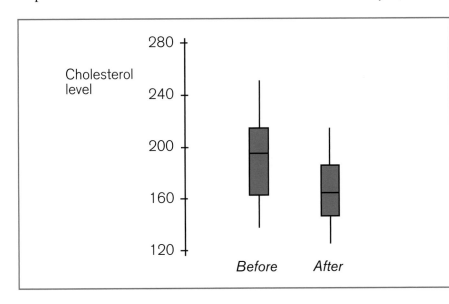

Fig. **3.7.**

Comparing *Before* and *After* box plots of the data from table 3.1

Sample data form independent samples if the measurements are obtained from two different groups of subjects selected from different randomizationss.

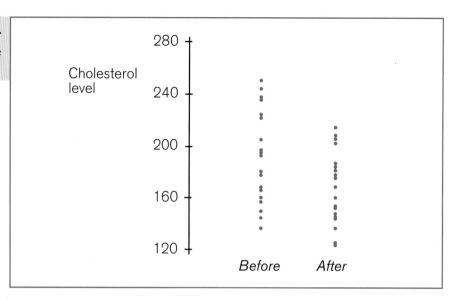

What are the chances that my cholesterol level will drop? By
about how much, if at all, can I expect it to drop?

To predict one's own chances of decreasing blood cholesterol by following the diet, one should begin by asking, "How many individuals'
levels decreased?" To gauge the size of one's own possible drop, one
should consider questions like "What was the largest decrease that anyone experienced?" and "What was the smallest observed improvement?"

The variable of most interest for the question about how much one
can expect one's own cholesterol level to drop is *change in level*. Values
for this variable can be found by subtracting: *Change = After – Before*. In
the context of the study, improvement is equivalent to negative change.
Thus, we might define the variable *Improvement* in such a way that
Improvement = Before – After, with positive values being "good." This is
the variable of interest used in figures 3.9 and 3.10, in which a box plot
and a dot plot, respectively, show the distribution of improvement in
the data.

Fig. **3.9.**

The data from table 3.1 in a box plot
showing improvement (decrease in
cholesterol level)

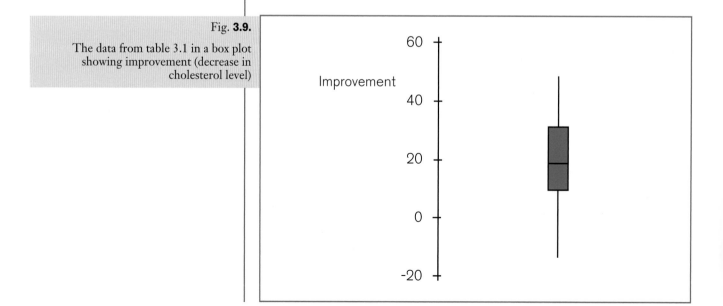

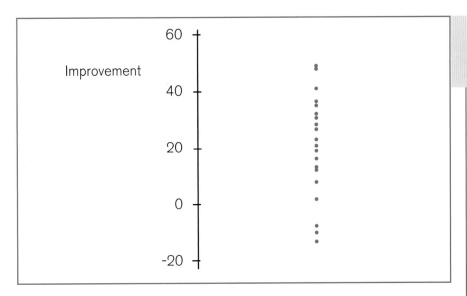

The center of these distributions is about 20, representing a typical improvement of about 20 mg/dL after the diet. Thus, this is the improvement that one could predict for a randomly selected individual *without* taking the individual's initial level into account.

It is worth noting that simply calculating the mean change does not give any sense of the variability around it. Are some changes very small and others very large? Are the changes small and clustered? The graphs make the variability in the data visible. In the box plot of improvement (fig. 3.9), the first and third quartiles appear to be about 10 mg/dL and 30 mg/dL, respectively. Thus, one can estimate the interquartile range, 30 – 10, or 20 mg/dL, of improvement that would be likely to occur as a result of the new diet. The dot plot in Figure 3.10 shows the spread and can be used to assess the probability that any particular individual will improve, since from the plot only 3 of the 24 volunteers have negative improvement scores.

How does my current cholesterol level affect what might happen if I use the diet? Can a doctor predict the change in cholesterol level for a patient who follows the diet?

The previous plots and discussions did not consider both the relationship between cholesterol levels before and after the diet and how the initial cholesterol level may affect any change in levels. Taking the differences between the "before" and "after" scores accounts for the fact that these are paired for the individuals from whom they were taken, but it does not account for their initial cholesterol levels. Comparing the two single-variable distributions allows one to see the initial cholesterol levels but hides the information on pairing. An approach that students typically use in investigating whether there is any relationship between "before" and "after" scores is to fit a line to the data and to base conclusions only on the slope of the resulting equation (see fig. 3.11).

The scatterplot shows that "after" and "before" data are related and that the relationship seems to be linear. Figure 3.11 shows the data modeled by the least squares regression line, but students may use other methods, depending on their background. The equation of the line in figure 3.11 can be expressed as *After* = 37 + 0.7(*Before*) after rounding the estimates of the slope and the *y*-intercept. This equation and its

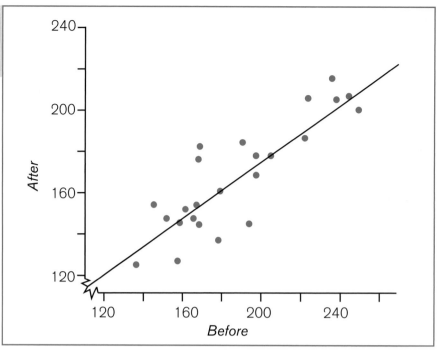

Linear regression is a method of estimating the equation of a line that describes the "average" relationship between two variables. It minimizes the sum of the squares of the vertical distances from the line to the observed data. Thus, the line is called the "least squares regression line." Statistical calculators and computer packages usually include a command for linear, or least squares, regression.

"For bivariate measurement data, [all students should] be able to display a scatterplot, describe its shape, and determine regression coefficients [and] regression equations." (NCTM 2000, p. 324)

corresponding line estimate the average "after" cholesterol levels for given "before" levels under this diet.

A good follow-up exercise for students who think that the slope determines whether individuals improve or not is to have them construct small counterexamples of their own. Have the students generate arbitrary pairs of points (*Before*, *After*), with each pair illustrating a different combination of slope and improvement in cholesterol level. Positive and negative slopes, some with absolute value greater than 1 and others with absolute value less than 1, should be matched both with decreased cholesterol levels and with increased levels. Such an exercise should help students discover how to use the scatterplot correctly to determine improvement.

Figure 3.12 illustrates another approach that uses a regression line. This approach examines the graph of the least squares regression line by applying the criterion *After < Before*. Note that the the figure omits individual points of the scatterplot to stress that at this point students can focus on the lines and their equations rather than on the data. Students may include the data points in their scatterplots but without using them, instead relying on associated equations.

The shaded region below the *After = Before* line represents improvement. Points above the line represent increased cholesterol levels, and points on the line represent no change. The least squares regression line describes the average "after" cholesterol levels under this diet for given "before" levels and can be used for predicting the expected "after" level for any given "before" level. Thus, the point at which the least squares line meets the *After = Before* line, about (135, 135), represents the point at which the predicted effect of the diet is "no change." One can now answer an important question: "What initial cholesterol levels are likely to become lower when patients follow the diet?" For initial cholesterol levels greater than 135 (the *Before*-coordinate of the intersection point), the diet is, on average, effective. For lower "before" levels, it is not.

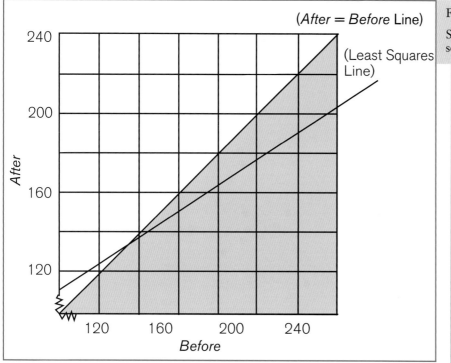

The regression line has a slope of 0.7, which is less than 1, the slope of the *After = Before* line. For "after" values greater than 135, the least squares regression line is below the *After = Before* line. Note that as the initial cholesterol levels increase, the vertical distance between the *After = Before* line and the least squares regression line also increases. Thus, the higher the initial cholesterol level, the greater the predicted improvement after the diet.

There are two important ideas in this graph. First, the *After = Before* line allows identification of improved cholesterol levels. Second, comparing the two lines reveals that the average amount of improvement resulting from the diet is related to the initial cholesterol level before the diet. People with higher initial levels tend to have larger decreases in levels, on average, than do people with lower initial values.

Does the diet work, and if so, for whom?

In the discussion of the first question ("Does the diet make a difference?"), we examined a scatterplot of the individual data points for "after" the diet versus "before," together with the *After = Before* line (fig. 3.2). Since *After = Before* represents no improvement, we might call this the "no-improvement line." As in the dot plot showing improvement in figure 3.10, figure 3.13 identifies individuals who improved and those who did not, this time by the shading of *After < Before*. The representation shows that improvement is related to the "before-diet" cholesterol level, much as figure 3.12 showed that higher "before" levels tended to predict greater improvement (as seen in greater distance between the lines).

At this stage, it is tempting to say that we know the effect of the diet on cholesterol. But suppose that we were doctors and we had a patient who said, "I have a cholesterol level of 250. What will my cholesterol level be if I adopt this diet?" Using our analysis, we could predict that it would be 37 + 0.7(250), or 212 mg/dL. However, the graph shown in

Fig. **3.13**

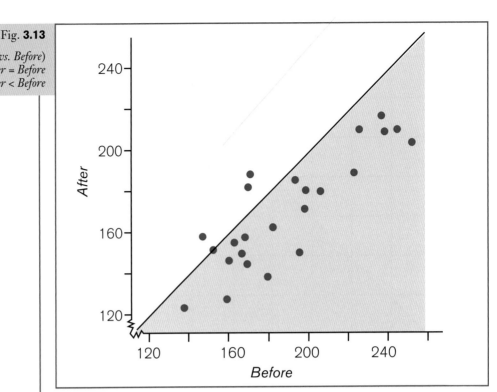

figure 3.13 would tell us that there is a *range* of plausible outcomes rather than a single value. There is still variation in the "after" cholesterol levels that diet does not account for.

In fact, for certain initial cholesterol levels, the range of likely "after-diet" cholesterol levels may include the no-improvement line. The proportion of points above the no-improvement line for similar initial cholesterol levels gives an estimate of the probability of increased cholesterol with the vegetarian diet for patients at that starting level. Thus, as doctors, we could predict the patient's level and a range in which that level might reasonably fall, and we could estimate the patient's chances for any improvement at all.

We could represent this last analysis on a scatterplot showing improvement versus cholesterol before the diet, as depicted in figure 3.14. Note that in contrast to fig. 3.13, which allowed us to observe improved cholesterol levels by comparing values to *After = Before*, fig. 3.14 shows improvement as the response variable, so here we concentrate on those values for which *Improvement* > 0.

Though the information obtained from this and some of the previous methods is equivalent, figure 3.14 has the advantage of using a variable that directly measures what is of most interest—*Improvement*. It makes clear that not only the *amount* of improvement but also the *probability* of improvement depends on the initial cholesterol level. The distance above the zero-line gives the amount of improvement. The proportion of points above the line for any given range of "before" values estimates the probability of improvement for a patient starting in that range. For example, about half of the subjects started at levels below 180, and about 25 percent of those failed to improve. By contrast, 100 percent of the subjects starting at levels above 180 showed improvement. Figure 3.14 makes clear that for all but three of the study participants, cholesterol levels improved. Moreover, those whose levels did not improve originally had low cholesterol levels.

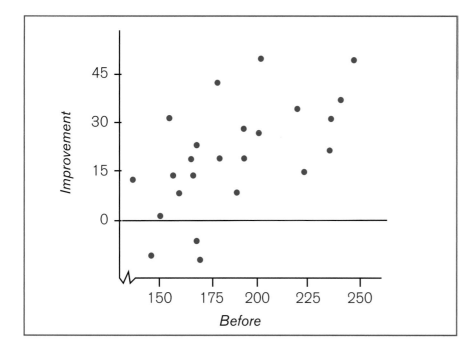

Treating improvement in the manner shown in figure 3.14—as a new variable created by returning to the original ordered pairs (*Before, After*) and subtracting (*Before – After*)—allows for additional analyses that can provide useful information for persons considering the dietary change. The relationship between improvement and "before" cholesterol levels is best shown in a plot of these variables (*Improvement* vs. *Before*), such as fig. 3.14 shows.

Is the observed decrease in cholesterol levels an improvement that is really the result of the new diet, or could so large a difference have occurred by chance variation?

The next activity, Simulating and Counting Successes, helps answer this question and shows how results can be analyzed and displayed.

"All students should … select and use appropriate statistical methods to analyze data." (NCTM 2000, p. 324)

Simulating and Counting Successes

Goal

To use a simple simulation to infer whether the improvements in the scenario Dietary Change and Cholesterol occurred by chance variation or as a result of other factors, such as a vegetarian diet

Materials

- A copy of the activity page for each student
- A copy of the activity sheet "Dietary Change and Cholesterol" for each student
- A fair coin for each student
- A statistical graphing utility or graph paper
- The Random Number Generator applet or another randomizing device

pp. 113, 117–18

Discussion

The questions in the activity provide a starting point for discussing statistical inference by using approaches similar to that in the preceding chapter, in which students dealt out cards to simulate the scenario Discrimination or Not? under random conditions, counted the cards, and analyzed results. Suppose an individual's cholesterol level were measured several times. Those measurements would not all be identical but would be likely to vary within some interval. If this is so, could the improvement seen in the two columns of data from the study represent just the natural variability among measurements of individual cholesterol levels? That is, could it be that the diet actually had no effect?

Note that three volunteers in the Dietary Change and Cholesterol study saw their cholesterol levels increase rather than decrease after following a vegetarian diet. If the diet actually had no effect on patients' cholesterol levels, students might expect that the increases—as well as the decreases—occurred purely randomly. The activity Simulating and Counting Successes gives them an opportunity to consider a number of additional issues related to this question in connection with the scenario:

- Suppose that the probability of cholesterol levels decreasing when people switched to a vegetarian diet really were 50 percent. Under this condition, what would be the probability that the cholesterol levels of at least 21 volunteers out of 24 would decrease in a study like that in the scenario Dietary Change and Cholesterol?
- What would your answer suggest about the diet?

Drawing Conclusions from Data

It may come as a surprise to students that the question that you originally considered with them in your discussion of the activity What Can You Know—How Can You Show?—"Does the diet make a difference?"—has not yet really been answered. The data obviously show improvement, but is the improvement enough to make the diet a desirable treatment? The important question is "Are the observed improvements possibly due to chance variation when no mean improvement exists, or are they actual mean improvements?" This is similar to the question posed in chapter 2 about Discrimination or Not? "Could the observed proportion of women who were recommended for promotion possibly have occurred by chance variation with no discrimination, or did inequitable treatment of women enter into the selection process?" The basic idea is that variability is inherent in the population and in the sampling process. Inference is the process of using the sample to make conclusions about the population being considered in the presence of this variation.

Two simulation models are discussed below. The first model examines whether improvements occurred *more often* than might be expected if the diet were not effective. The second one examines whether improvements were of *greater size* than might be expected if the diet were not effective. The two models differ in their choice of sample statistic. The first uses "number of improved volunteers"; the second uses "mean improvement." Both models can be represented with simulations, and the underlying randomization processes in the two simulation models are identical. The activity Simulating and Counting Sucesses illustrates the first model; the second model has no example in an activity here.

Binomial inference. The question "Will I improve under the new diet?" leads to an analysis of success-failure outcomes. A decrease in cholesterol constitutes success; increased cholesterol represents failure. Thus, the statistic of interest can be thought of as a particular quantity—"number who improve." If the diet actually had no effect, we would expect about half the outcomes to be successes and about half to be failures, since each outcome would be due only to chance variation, not to any effect of the diet on cholesterol level. Hence, if the diet produced no improvement, we would expect the probability of improvement to be simply $p = .5$, which is the null hypothesis. So in a set of 24 volunteers, we would expect about 12 to improve. In the case that the diet really did not work, then, what would the chance be that as many as 21 would improve?

The activity Simulating and Counting Successes enables students to investigate this question by carrying out a simple simulation that is similar to the one used in the activity Simulating the Case in chapter 2. The variable for students to observe here is *Number who improve*. A simulated sampling distribution of this statistic generated in sets of 24 simulated volunteers, with each volunteer having a fifty-fifty chance of improving, can be used to assess the likelihood that at least 21 volunteers would improve if the diet had no effect whatsoever.

The activity page for Simulating and Counting Successes summarizes the elements that are essential in a simulation. A number of mechanisms are appropriate for valid simulations. The activity uses a coin, but students might choose to use dice, with even numbers representing success and odd numbers representing failure. Or students might design their own simulation.

After students have tried the simulation enough times to understand the connection between its structure and the study's features and to know how to use the simulation's results to estimate the probability of

The Lowering Cholesterol— Counting Successes applet on the CD-ROM can assist students in carrying out the simulation.

"All students should … use simulations to explore the variability of sample statistics from a known population and to construct sampling distributions."
(NCTM 2000, p. 401).

at least 21 successes, they can switch to calculators or computers to generate enough trials (say, 300 or more) to give stable distributions. One such simulated sampling distribution is illustrated in figure 3.15. Note that in this sampling distribution, the estimated probability of having 21 or more successes is 0 when the diet has no effect and $p = .5$. Thus, there is strong evidence that 21 successes did not occur as a result of chance variation and that the diet did have an effect on cholesterol level.

Fig. 3.15.

A sampling distribution of simulations of the scenario Dietary Change and Cholesterol

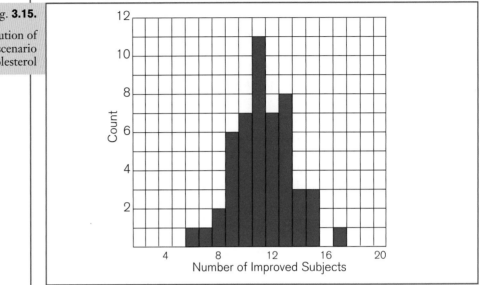

Drawing conclusions from sample mean improvement. The previous model considers whether improvements occurred more often than might be expected if the diet had no effect—a question that calls for a success-failure characterization. The second model offers another approach. Rather than counting subjects who improve, we could consider the actual *amounts* of improvement. If the diet were not effective, we would assume that the mean improvement would be 0. In this case, the null hypothesis would be that the mean improvement is 0.

A competing null hypothesis, that the *median* improvement is 0, is equivalent to the null hypothesis $p = .5$, tested in the previous simulation. To understand that this is true, note that a subject is considered a success in that setting *if and only if* the subject's change is positive (i.e., his or her second cholesterol level is lower than the first). But the median improvement is defined as the value for which the probability of positive change is .5. So testing that the median is equal to 0 is the same as testing that the probability of being above 0 (i.e., of being a success) is .5.

To test this new null hypothesis—that the diet is not effective and the mean improvement is 0—we need a simulation that uses the numbers in the original data to generate a simulated sampling distribution of the sample mean improvement under the null hypothesis. Under the assumption that the diet was not effective, we may think of the cholesterol levels in the table as assigned to "before" and "after" randomly.

For example, the first volunteer had cholesterol levels of 195 and 146. Over the course of a month, a person's cholesterol level may fluctuate. Could the fact that this volunteer's "before" level was 195 and his or her "after" level was 146 have been due to that fluctuation? Could the measurements, through normal variation, have been reversed? That is, although we must continue to regard the values as paired since they came from a single subject, could we think of the value for "before" and that for "after" as due only to chance variation over time, rather than to dietary change, for that individual? The same is true of each pair of measurements. Let's assume that each pair represents only random fluctuations and that the assignment to "before" and "after" is due only to chance variation, not change in cholesterol level as a result of the diet.

These assumptions allow us to create a simulation that randomizes the assignment of each person's levels to "before" and "after" and lets us compute the improvement for each person and then record the sample mean improvement for the set of 24 pairs. We must repeat the simulation enough times to generate a simulated sampling distribution of sample mean improvements under the null hypothesis of no change in cholesterol as a result of the diet. Then we can locate the observed sample mean improvement (19.5) within that simulated sampling distribution.

For example, we might toss a coin 24 times to carry out one round of our randomization, with heads meaning that we will label the first number for a given patient "before" and tails meaning that we will label the first number "after." Then the string of coin tosses HTTHT... would allow us to create table 3.2, which shows the results of this round.

The mean improvement for this set of 24 simulated individual improvement scores is –0.125. With repeated simulations, we could build the sampling distribution of mean improvement.

Table 3.2
Simulated (Randomized) "Before" and "After" Cholesterol Levels

Original Before	Original After	Original Improvement	H/T	Random Before	Random After	Improvement
195	146	49	H	195	146	49
145	155	–10	T	155	145	10
205	178	27	T	178	205	–27
159	146	13	H	159	146	13
244	208	36	T	208	244	–36
166	147	19	H	166	147	19
250	202	48	H	250	202	48
236	215	21	T	215	236	–21
192	184	8	H	192	184	8
224	208	16	T	208	224	–16
238	206	32	H	238	206	32
197	169	–28	T	169	197	–28
169	182	–13	H	169	182	–13
158	127	31	T	127	158	–31
151	149	2	T	149	151	–2
197	178	19	T	178	197	–19
180	161	19	H	180	161	19
222	187	35	H	222	187	35
168	176	–8	T	176	168	8
168	145	23	H	168	145	23
167	154	13	T	154	167	–13
161	153	8	T	153	161	–8
178	137	41	T	137	178	–41
137	125	–12	T	125	137	–12

For another discussion of the use of simulation for randomization tests–that is, tests of the likelihood of particular events occurring under random conditions–see Barbella, Denby, and Landwehr (1990) on the accompanying CD-ROM. (Good [1999] provides a more technical treatment.)

"All students should ... use simulations to explore the variability of sample statistics from a known population and to construct sampling distributions."
(NCTM 2000 p. 324)

The Lowering Cholesterol— Mean Improvement applet on the accompanying CD-ROM can assist students in generating the simulated sampling distributions.

Researchers would commit a type I error by concluding that a vegetarian diet lowered blood cholesterol if the diet did not have an effect on the patients' average cholesterol levels. Understanding Type I and Type II Errors on the CD-ROM discusses errors of both types in relation to the scenario.

In practice, this simulation may be accomplished by multiplying each actual improvement value by a randomly generated 1 or –1 and then computing the mean of the resulting list. In the example given, then, the string HTTHT would be replaced by 1, –1, –1, 1, –1, which would be used to multiply the actual improvement scores (49, –10, 27, 13, 36) of the corresponding raw data, yielding the same simulated results more quickly. Such calculations are particularly simple on spreadsheets or calculators that permit list operations.

Note that the randomizing process in this model is exactly the same as that used earlier for the binomial simulation. In the binomial case, the 24 coin tosses classified each person's outcome only as success or failure, and the resulting successes were counted. Here, the coin tosses classify the corresponding improvement score as positive or negative, and the resulting scores are averaged.

Conclusion

Statistics is commonly defined as the art of finding information in the presence of variability. Variability is a feature of all data. It needs to be noticed, described, and interpreted in each new setting. Examining the scenario Diet and Cholesterol involved students in making a decision about a set of bivariate data where there is variability and uncertainty. Initially, their focus was on what questions to ask and what graphical displays might help in finding answers to those questions. They needed to consider questions such as the following:

- What are the variables of interest?
- Is there a relationship between the variables?
- What feature of the relationship is important?
- What graphs should you draw and what features of the graphs should you examine?

Ultimately, the examination of the scenario led to the formulation of this important question:

- Could the outcome of the study (either the number of volunteers whose cholesterol level improved or their mean improvement) reasonably be attributed to chance variation, or would the observed outcome be more likely to be due to some other effect—in this case, the diet?

The most important question to ask in choosing an analysis of data is always "Why?" It is not enough just to report the conclusions reached. What is valuable is to enable others to reach them, too.

In statistics, decisions are only as good as the data and the analyses. This chapter focused on the analysis. What affects the quality of the data? Does it matter how the sampling *units* (in this problem, the volunteers) are selected? Does it matter how they are "treated" after they are selected? The next chapter examines such issues and others related to data production.

NAVIGATIONS SERIES

GRADES 9–12

NAVIGATING *through* DATA ANALYSIS

Chapter 4
Designing Studies

The first chapters of this book considered the roles that variability, sampling, and formulating questions all play in making decisions with data. This chapter focuses on how to design studies. The process begins by recognizing the effect that the design has on the conclusions that can ultimately be drawn from the results. The first part of the chapter focuses on evaluating designs of studies reported in the media. The second part addresses the components of a well-designed study, and the final part shows how to engage students in designing a study of their own and critically evaluating its design.

Evaluating the Designs of Research Reports

The media often provide conflicting reports on important issues. For example, in the last several years, news reports have cited research studies that supposedly show that cell phones cause cancer (e.g., Leske 2001) as well as other studies that supposedly show that they do not (e.g., Muscat et al. 2000). The studies are attempting to answer the same—or at least what appears to be the same—question. How could they reach conclusions that are so different? Which conclusion is correct? Answering these questions involves understanding principles of research design. The activity What Does This Study Do? provides an opportunity for students to begin to think about the design of studies.

What Does This Study Do?

Goals:

- Understand components of study design
- Think critically about a study or reported study
- Consider the potential problems of a study that is based on insufficient data
- Consider how, or if, the results of a study apply in general

Materials

- A copy of the activity pages for each student

Discussion

In What Does This Study Do? students examine summaries of three studies of effects of cell phone use. The first was reported by the researchers themselves in the *Journal of the American Medical Association* (Muscat et al. 2000), and the other two were reported in news accounts in the London *Sunday Times* (Leske 2001) and the *Australian* (Fist 1997). Recognizing what types of studies these are supplies students with an important key to understanding the different conclusions in the accounts. Both the study reported by Muscat and others and that reported by Leske were *observational studies*—that is, they involved neither a random selection of subjects for the study nor a random assignment of treatments (here, cell phone use or no cell phone use). The study reported by Fist was an *experiment*, in which subjects were randomly assigned to treatments.

As reported in the *Journal of the American Medical Association*, the study by Muscat et al. (study 1 in the activity), found no significant risk from short-term use of cell phones with analog signals. The study compared the cell phone habits of 469 people with brain cancer to those of 422 healthy people matched by age, gender, and other characteristics. The cell phone use of the group with cancer averaged 2.5 hours per month, compared with the control group's 2.2 hours per month. Although the difference of 0.3 hours on the cell phone each month was not significant, the average time on the cell phone of the group with cancer was higher. Does this study imply that the likelihood of getting brain cancer increases as analog cell phone use increases? Also, cell phones are typically digital instead of analog. Would the results of this study apply to digital cell phones as well?

The study reported by Leske (study 2 in the activity) compared a group made up of 118 people with an eye cancer (uveal melanoma) with a control group of 475 people without the disease. The group with cancer had a much higher rate of cell phone use than the control group. In the report of this study, there was no indication whether either the choice of the control group or the statistical analysis took into account possible group differences in age, gender, or other characteristics that might have affected the results. These are important considerations because a factor such as age—not cell phone use—could be the primary contributor to the cancer.

pp. 119–20

"All students should … understand the differences among various kinds of studies and which types of inferences can legitimately be drawn from each."
(NCTM 2000, p. 324)

Navigating through Data Analysis in Grades 9–12

In these two observational studies, people with and without cancer were recruited, rather than randomly selected, for the study, and the cell phone use of the two groups was compared. If the studies showed, as the second one did, a significant difference in cell phone use by the two groups, they could, at most, point to an *association* between cancer and cell phone use for those who participated.

To establish a cause-and-effect relationship between cancer in people and cell phone use, researchers must conduct an experiment, in which a random assignment of treatments is imposed on human subjects and their responses are measured. Developing an experiment to investigate whether a cause-and-effect relationship exists between cancer and cell phone use is not a simple task. Such a study necessarily involves both human subjects and their random assignment to cell phone use. Because researchers would not be able to guarantee that people would not be harmed by such an experiment, most institutions would not allow such studies. In circumstances where a significant possibility of harm exists, researchers often design experiments that use animals. For example, an experiment investigating the link between cell phone use and tumors in mice was reported several years ago by Fist (1997) in the *Australian* (study 3 in What Does This Study Do?).

This study, conducted by researchers at the Royal Adelaide Hospital, exposed one hundred mice to cell phone radiation for two half-hour periods each day for eighteen months and fitted another one hundred mice with the same type of antennas, which never had the power turned on. The number of tumors was the reponse variable measured for each mouse. Tumor rate, the statistic computed from the response variable, was twice as high in the exposed group as in the unexposed group. Because conditions were controlled for both groups and only the treatments (the presence or absence of cell phone radiation) differed between the two groups, the increased tumor rate in the exposed group was attributed to the radiation.

Does this study imply that cell phone use causes cancer in humans? Not necessarily. Mice are smaller than humans, and cell phone radiation may have a greater impact on them than it would on humans receiving the same dosage. Confined laboratory conditions could have intensified the mice's exposure to the cell phone radiation. Moreover, the mice were genetically modified laboratory mice; humans lack the gene that had been added to the mice to increase their likelihood of developing cancer quickly. These differences could prevent the results for mice from applying to humans.

An experiment on humans would be needed to show whether cell phone radiation increases the risk of cancer. Yet, as noted earlier, finding people to participate in a potentially dangerous study is difficult or impossible, if not potentially unethical. However, thinking about such a study helps students identify the numerous ethical issues that researchers must address before undertaking any such experiment. These ethical issues provide clues about why observational studies are more common than experiments when human subjects are involved.

Throughout high school, students should increase their understanding of the basic differences among observational studies, sample surveys, and experiments, as well as the types of conclusions that can be drawn from them. All three types of studies benefit from careful design. Their distinguishing features are the selection of "units"—the smallest

Researcher Michael Repacholi et al. (1997) probed the connection between cell phone radiation and tumors in mice in a study at the Royal Adelaide Hospital in Australia. Accounts of their study have appeared in the popular press (e.g., Fist 1997), and their own report of their study was published in the journal *Radiation Research*.

Some books on statistics treat sample surveys as observational studies; however, for clarity, we have distinguished three categories of studies:

- observational studies
- sample surveys
- experiments

"All students should ... formulate questions that can be addressed with data and collect, organize, and display relevant data to answer them."

(NCTM 2001, p. 324)

entities (either individuals or groups) drawn from the population of interest—that the researchers plan to study—and the assignment of those units to treatments.

Observational studies make no random selection of units from the populations and no random assignment of those units to treatments. Instead, as their name implies, they simply observe the characteristics of a group of units from one or more existing populations. The first two studies—Muscat et al. (2000) and Leske (2001)—were both observational studies. The units were people, with and without cancer, who were willing to participate. Only *associations*—or the lack of any association—between cell phone use and cancer can be identified and only for the people in those studies.

Sample surveys select units randomly from the population of interest for inclusion in the study but make no random assignment of them to treatments. The results of a sample survey can be used to identify associations among variables for the population from which the units were randomly selected—not just for the people in the study, as in an observational study.

Experiments assign all units randomly to treatments. They allow researchers to draw cause-and-effect conclusions for either the study's units (if the experimental units were not randomly selected) or the population (if the experimental units were randomly selected).

Concepts of Design for Studies

Before designing a study, a researcher should—

- develop a clear statement of the study's objectives;
- investigate prior, related research; and
- determine available resources for the study.

Design is important to *all* types of studies, of course, but most of the discussion that follows focuses on experiments.

In establishing the objectives for a study, one challenge that the researcher faces is to define its scope clearly. It is often tempting to ask broad questions that are difficult, if not impossible, to answer with a single study. Investigators generally make better progress when they address smaller facets of the problem through a series of studies, building to the larger question. Although it is good to have the ultimate question in mind, it is also important to limit the objectives of a particular study to what the study can actually accomplish with the available resources. A cautionary note is in order: Sometimes when researchers break a problem into smaller parts, study the parts, and then put them back together, expecting to answer the original question, they find that the sum of the parts is not really the anticipated whole. Researchers must take care when they separate a problem into component parts to ensure that this will not happen.

Once investigators have clearly established their research objectives, it is important for them to learn what work relating to these objectives has already been completed. A literature search allows them to determine how their proposed work relates to what has been done by others. Sometimes they discover that other researchers have fully addressed the objectives of their proposed study, and a new study is not needed. Other times, the literature reveals that a new study, either with the objectives proposed or with modified ones, could yield more information. Typically, an iterative process—a series of reconsiderations and refinements—accompanies the literature review and the establishment of a study's objectives. Each iteration affects the next, and all are tempered by the availability of resources.

After the investigators have set the study's objectives and have reviewed the relevant literature (if any), they can begin the design phase. During this phase, they should—

- specify the response variables and indicate how they propose to measure them;

- determine the experimental units—the individuals or groups that they will randomly assign to treatments;

- define the *scope of inference*—the population to which the experiment's results can be legitimately applied;

- choose the treatments that they will compare;

- identify major sources of variation;

- determine potential errors and the level of acceptability of those errors;

- select an experimental design; and

- outline a plan for analyzing the data collected.

A fundamental decision that researchers make in designing an experiment is the choice of the experimental unit—the smallest object to which a treatment will be applied at random. In a study in which people are given treatments, the experimental unit might be a single individual or a set of individuals, depending on whether individual people or groups of people are randomly assigned—*randomized*—to treatments. In the mice study, individual mice were randomly assigned to the treatments (exposure to cell phone radiation or no exposure to cell phone radiation), making a mouse the experimental unit. Thus, there were one hundred *replications* (mice) for each treatment.

Suppose that instead of randomizing individual mice to cell phone exposure or no cell phone exposure, the researchers had used a different randomization process. Imagine that they placed one hundred mice in one cage and one hundred in another and then randomly selected a cage to receive the cell phone radiation. In such a case, because the cages were randomly assigned treatments, the experimental unit would be a cage of one hundred mice and not an individual mouse. All the mice in the study would not respond in the same way. For two mice in the same cage, differences between responses could be due to differences between the two mice. For two mice in different cages, however, differences between responses could be due to differences between the mice, differences between the cages (perhaps one got more light or heat than the other, for example), or differences between the treatments.

For this example, if cages of mice rather than individual mice were assigned to treatments, there could be only one cage of mice (experimental unit) per treatment, and hence, there would be no replication—that is, assignment of more than one experimental unit to each treatment. Without replication, the researchers would have no measure of variability among cages receiving the same treatment. Without a

"All students should … know the characteristics of well-designed studies." (NCTM 2000, p. 324)

measure of variability among cages, it would not be possible to separate differences between cages from differences between treatments. As a result, the researchers could make no valid statistical inference. (Note that in the actual study, individual mice were the experimental units, ensuring replication.)

The manner in which the experimental units are chosen for inclusion in a study determines the scope of inference, defined earlier as the population to which the experiment's results can reasonably be applied. If the experimental units are randomly selected from a population, then inference can be drawn to that population. However, if investigators make no random selection of experimental units, then they can draw inference only to those units used in the study. Figure 4.1 shows the scope of statistical inferences in observational studies, sample surveys, and experiments.

As noted earlier, the key to distinguishing among observational studies (cell D in fig. 4.1), sample surveys (cell B), and experiments (cells A and C) is determining whether or not there was a random selection of units and/or a random assignment of them to treatments. Suppose a study were designed to compare the absorbency of brands of paper

Fig. 4.1.

Statistical inferences permitted by study designs. (Adapted from Ramsey and Schafer 1996, p. 9.)

		Allocation of Units to Treatment Groups		Statistical Inference Permitted
		At Random	Not at Random	
Selection of Units	At Random	A. **Experiment with broad scope of inference.** A random sample of units is selected from one population; units are then randomly assigned to treatments.	B. **Sample survey.** Random samples of units are selected, but there is no randomization of these units to treatment.	Inference can be drawn to the population.
	Not at Random	C. **Experiment with narrow scope of inference.** A group of units is found (not selected randomly); the units are then randomly assigned to treatment groups.	D. **Observational study.** Available units are collected from distinct groups and observed (no random selection or random assignment to treatments).	Inference is limited to the units included in the study.
Statistical Inference Permitted		Causal inference can be drawn.	Associations can be made, but no causal inference can be drawn.	

towels. Absorbency could be measured in several ways. One possibility would be to use an eyedropper to release a drop of water onto the center of a suspended paper towel sheet every five seconds until the sheet dripped some water. The response variable would be the number of drops of water released onto the towel before it dripped. The unit of study would be a paper towel sheet. How the researchers chose the paper towel sheet would then be an important question.

Suppose a few different brands of paper towels were available for the study. If the study used one roll of each brand of paper towel and tested the first few sheets off each roll, then no random selection of units would occur at all. The sheets could be considered a convenience sample. The sheets—the units in the study—could not be randomly assigned to a brand, which represents a treatment here, because the brand is an inherent property of the sheet. Because the study used neither random selection of units nor random assignment of treatments, it would be an observational study (cell D), and associations might be found between the brands and the absorbency only of the particular sheets used in this study—not other sheets from these brands.

If the rolls of paper towels for each brand in the study were randomly selected from all the rolls available in a city, and then the researchers chose a sheet at random from each selected roll to have the water dripped on it, the study would be a sample survey. On the basis of the results of such a study, the researchers could make associations among paper towel brands and absorbency, at least for the paper towel rolls in the city. In this case, as in the case of the observational study, no causal inference could be made, because the study made no random assignment of the units to treatments (brands).

As a contrasting example, consider a study in which the researchers wanted to compare the weight gain of puppies eating one brand of dog food with that of puppies eating a different brand. Suppose that, with a puppy as the experimental unit and the dog food brand as the treatment, they randomly assigned each puppy to a brand of dog food. This would clearly be an experiment (cell A or C), and the researchers could use the results to draw causal inference.

The scope of inference would depend on the manner in which the researchers selected puppies for inclusion in the study. If the puppies were chosen completely at random from the population of all puppies and randomly assigned to treatments, then this study would be an experiment with the broadest possible scope of inference (cell A). However, most studies of this type obtain puppies in the easiest manner. One readily available source is the local animal shelter. If the researchers used puppies that happened to be on hand at the shelter, they would not be randomly selecting the puppies from the general population of puppies. Consequently, any inference could be drawn only to the puppies in the study. As a result, the study would be an experiment with a limited scope of inference (cell C).

If the researchers conducting the experiment with puppies from the shelter wanted to justify broadening the scope of inference, they might contend that these puppies could be assumed to make up a random sample from some larger population of puppies. This assumption would permit recommendations of one brand of dog food over the other on the basis of experimental results about the puppies' weight gains. However, the validity of broadening the scope of inference from the puppies in the study to those in the larger population could be called into question. One could argue that puppies that have spent any time in an animal shelter would respond differently from other puppies to the two dog food brands. If that argument were accepted, the results could not be extended to the general population of puppies. Nothing in the experiment as described would provide a foundation for arguing otherwise.

In designing the puppy study, the researchers would also need to consider how they would select dog food from each brand. If they bought only one bag of each brand and fed all puppies in the study from one bag or the other, they could draw inferences only about the two bags of dog food that they used in the study.

Large variation in a response variable can make it difficult for researchers to detect differences in treatments. By identifying major sources of anticipated variation before the study, a researcher can either limit the scope of inference by controlling this variation or try to choose a design that accounts for it. In the hypothetical dog food study, the researchers used a *two-group design*, in which they randomly assigned puppies to dog food brands, thus creating two groups.

The size of the puppies could be a major potential source of variation for the investigators to control or account for in such a study; a toy poodle, for example, could be expected to gain less weight than an Irish setter because of general differences in size between the breeds. The researchers could choose to consider only poodles or only Irish setters, thereby reducing the scope of inference as well as the variability. In addition, to account for this variation through the design of their study, they could use a *paired design*, with two puppies from the same breed composing a pair. Within each pair, one puppy would be randomly chosen and fed the first brand of dog food; the other puppy would be fed the other brand. This process would be continued for each pair of puppies in the study. Because puppies within a pair were of the same breed, differences in their responses would reflect differences between puppies and differences between treatments (dog food brands) but not differences among breeds. Thus, the variability among breeds would be accounted for in the study's design, allowing the researchers to detect smaller differences in puppies' responses to brands of dog food.

Observed differences in treatment means can be of *practical importance* without being statistically significant. That is, they can be large enough for researchers to think that they are important, even though the probability of observing differences that are this large or larger when there are no differences in treatment means is fairly high. Differences in treatment means can also be statistically significant but not practically important. In the cell phone experiment on mice, the tumor rate was twice as high for mice exposed to cell phone radiation as for unexposed mice. This result was considered to have practical importance as well as statistical significance. However, if only ten mice (and not a hundred) had received each treatment, then a tumor rate that was twice as high for the cell phone radiation group might still have practical importance, but it would be unlikely to be statistically significant. In contrast, if a thousand or more mice had received each treatment, then arbitrarily small differences in the tumor rate could be found to be statistically significant even if they had no practical importance. By

Recall that an event is statistically significant if it is rare enough that it would be unlikely to occur by chance if the null hypothesis were true. That is, the likelihood of its occurrence must be small—traditionally, less than 5 percent. The choice of 5 percent (or 1 or 10 or some other percent) depends on the application. (See chapter 2 for discussions of *p*-value and significance level.)

"All students should … develop and evaluate inferences and predictions that are based on data."

(NCTM 2000, p. 324)

adjusting the sample size, the researcher can begin to equate practical and statistical significance.

As discussed in chapter 2, there are two kinds of errors that researchers can make in drawing conclusions from their studies. The first—a type I error—is to reject a true null hypothesis. The second—an error of type II—is to fail to reject a false null hypothesis. The following example can help students understand these errors.

In 1975, the Environmental Protection Agency (EPA) set the maximum level of arsenic in drinking water at 50 parts per billion (ppb). Over the years since that time, researchers have questioned whether this standard was too high, allowing too much arsenic in water systems, and studies have been conducted. In these studies, the null hypothesis would be that drinking water with 50 ppb of arsenic is safe. The alternative hypothesis would be that 50 ppb of arsenic is not safe. Hence, two decisions would be possible. The first would be that the null hypothesis is incorrect, and a level of 50 ppb of arsenic in drinking water is not safe. If the EPA accepted this conclusion, many cities would have to update their systems to remove arsenic from their water until it was below a new acceptable limit.

Suppose that researchers concluded that 50 ppb was an unsafe standard for arsenic in drinking water and this conclusion was incorrect. This would mean that the researchers had rejected a true null hypothesis, thereby committing a type I error. This error could cause cities to go to great and needless expense in reducing arsenic levels when a standard of 50 ppb in fact set a safe level. To avoid such a result, statistical analyses customarily make sure that the probability of this type of error is low.

The researchers' other possible decision would be that there was not enough evidence to reject the null hypothesis that arsenic levels of 50 ppb are safe. If the EPA adopted this result, cities would not have to update their water systems. However, if this decision turned out to be incorrect, researchers would have failed to reject a null hypothesis that was false, thereby committing a type II error. As a result, people's health would suffer because the arsenic levels in some drinking water would be too high to be safe.

To decrease the probability of this second type of error occurring, researchers must increase either the probability of the first type of error or the size of their sample. Typically, costs constrain what can be done in a research study. Increasing the sample size usually increases the costs. The relative seriousness of the different kinds of errors, as measured by their consequences, and the practical limits on sample size are important considerations in studies. Researchers need to decide early in the design process what might be acceptable.

Once all phases of design have been considered, the actual design of an experiment generally follows. Sketching out an analysis for the proposed experimental design provides researchers with another check on the design phase of the study. If the researcher does not know how to undertake an appropriate analysis of a proposed design, then the design phase is not complete.

Evaluating Examples of Studies

Students should apply the concepts of design discussed in the preceding section when reviewing research studies. They can look again at

In February 2002, a new EPA rule about arsenic in drinking water became effective. It requires states to comply with a new standard of no more than 10 ppb by 23 January 2006.

 See Understanding Type I and Type II Errors on the accompanying CD-ROM for additional discussion.

the studies presented earlier in the book. For example, in chapter 2, the treatments in Discrimination or Not? were files labeled "male" and "female." The response variable was the recommendation. Because the recommendation was either yes or no, the response variable was categorical rather than numerical. The supervisors were randomly assigned to "male" or "female" files (treatments), so the study was clearly an experiment (cell A or C of fig. 4.1). Because we do not know how the supervisors were chosen, we must limit our inference to the supervisors participating in the study (cell C). An implicit assumption in the analysis is that the bank supervisors did not know that others had the same file, so their decisions were independent.

In Dietary Change and Cholesterol in chapter 3, each "before" measurement was paired with an "after" measurement from the same individual. Such pairing permits the variability among people to be quantified and separated from the variability within a person. Because the variability among people is usually greater than that within a person, pairing allows smaller differences in the mean response (average change in cholesterol level) to be declared significant than would have been permissible without the pairing.

Students can also determine the scope of inference for the Dietary Change and Cholesterol study. The participants were hospital employees who volunteered for the study. Because these subjects were self-selected, they cannot be considered to compose a random sample of all employees in the hospital. Some employees chose not to participate. It may be that those who chose to participate have a systematically different response to a change in diet from those who did not. Or perhaps this is not so. It is not possible to know. Because the subjects were not randomly selected from all employees at the hospital, the study's scope of inference cannot be to all people, or even all the hospital employees. In fact, it is limited to the twenty-four individuals who participated in the study.

Designing a Study

In the activity How Fast Do They Melt in Your Mouth? students use components of research design in an experiment that addresses the question "Do real semisweet chocolate chips melt faster in your mouth than chocolate-flavored chips?"

How Fast Do They Melt in Your Mouth?

Goals

- Use research design components and concepts to design an experiment
- Think critically about a study
- Consider how, or if, the results of a study apply in general

Materials and Equipment

- A copy of the activity pages for each student
- A random number table, the Random Number Generator applet, or a calculator or computer for generating random numbers

p. 121–22

Discussion

In this activity, students consider the following scenario:

> A baker is trying to decide what chips to use in chocolate chip cookies that he plans to offer for sale. A salesman has told him that chocolate-flavored chips are "just like" real semisweet chocolate chips but cost much less. After thinking about it, the baker decides that what delights people most about chocolate chip cookies is the way the chocolate just melts in their mouths. If the baker can be sure that there is no difference between real semisweet and chocolate-flavored chips in this respect, then he will use chocolate-flavored chips in his cookies.

To have students conduct this study or any similar one, initiate a class discussion to give the students direction and focus, especially in the design stage.

The students need to agree on the response variable to observe and the conditions for each observation. The time from when a person puts a chip into his or her mouth until it melts is an appropriate response variable. Conditions may easily vary from person to person. Some people may suck on the chip, causing it to melt more rapidly than if they just hold it in one place on their tongue. Some may have been chewing gum immediately before putting the chip in their mouths. Others may have been drinking a cold beverage. These activities could affect how rapidly the chocolate chips melt. If investigators do nothing—or can do nothing—to eliminate these effects, then the variability in the observed times that it takes for the chips to melt in the students' mouths might be—probably will be, in fact—greater than it would if such effects were controlled. As the variability increases, the observed differences in the average melting times for the two types of chips must be greater before the researcher can be comfortable about saying that the differences are real and cannot be explained by the chance variation alone.

This activity was suggested by George Milliken, Kansas State University, in a personal communication to the authors.

Students should decide on a consistent set of conditions in which to make the observations. Controlling the conditions decreases both the scope of inference and the observed variability. For example, suppose that students decided that before putting the chips in the subjects' mouths, the subjects would drink a five-ounce cup of water at a particular temperature. Also suppose that the subjects were not to suck on the chips while they melted. Then inference could be drawn about the mean difference in the time for real semisweet and chocolate-flavored chips to melt in a mouth under these conditions but not under other conditions. At the same time, by controlling the experimental conditions, the students would ensure that the variability in the observed times would be less than it would be otherwise.

Here, in addition to assessing whether a random selection of experimental units and a random assignment of treatments have occurred, we must identify the study's goal in order to determine the type of study. The goal is to assess whether one type of chocolate chip melts faster *in a person's mouth*, on average, than another type of chip. Having the chips melt in people's mouths is a critical part of meeting the study's objective. People are the experimental units. Because a random assignment of chip type (treatment) is made to each participant (experimental unit), the study is an experiment—not a sample survey or an observational study. A significant difference in the observed average melting times can be attributed to differences in the real semisweet chocolate and chocolate-flavored chips of the particular brands studied. If the goal were simply to determine which type of chip melts faster, the chips could be melted by people, in ovens, or through some other means. In such a case, much as in the paper towel study discussed earlier, the study would be either a sample survey or an observational study, depending on whether or not there was a random selection of chips.

In the experiment that the students are designing, study participants will be limited to the students in the classroom, so inference will also be restricted to the students in the classroom. This scope may not be entirely satisfying to the students, and they may want to conclude that the results can be applied more generally. Perhaps this conclusion would be valid. However, if someone were to challenge it by claiming that something unusual about the students in the class made the results unusual, the study would offer no basis for refuting that claim.

When considering the populations of chocolate chips (chocolate-flavored and real semisweet chocolate), it is not reasonable to think that a classroom experiment could take a random sample of all possible chocolate chips. In fact, a bag of one brand of real semisweet chocolate chips and a bag of one brand of chocolate-flavored chips will probably be all the chips that are available. The students will select chips from each bag, but they might not do so randomly. Therefore, strictly speaking, inference can be drawn only to the chips in the study. Students will probably not find this satisfactory, since these chips will no longer exist when the study is completed.

Because of the uniformity in food processing, students may be comfortable about drawing inference to the chips in the bag, the chips of the two brands in the grocery store from which the bags were purchased, the chips of the two brands that come from the same processing plant, or maybe all chips of those two brands. However, if anyone later contended that the chips tested were unique, the lack of a random

"All students should ... know ... the role of randomization in surveys and experiments."
(NCTM 2000, p. 324)

Navigating through Data Analysis in Grades 9–12

sample of chips would counter any argument that this was not the case. Students who suggest that each chip should be randomly selected from a bag that was randomly selected from a randomly selected grocery store will really have grasped the idea of the importance of a random sample when drawing inference about a population.

Analyzing Results

The activity How Fast Do They Melt in Your Mouth? can help students gain a better understanding of the differences in two-group and paired designs and develop their understanding of design concepts. To make sure that your students explore these differences, you can randomly divide the class into three equal groups. Two of the groups can serve as experimental units in a two-group experiment, and the third can be the experimental units for a paired experiment.

For the two-group experiment, the students serving as experimental units should be randomly assigned to treatments—half to real semisweet chips and half to chocolate-flavored chips. (In general, it is not necessary to have the same number in each group.) The randomization could be done in several different ways. The assignment of students to groups could be made with random numbers from a random number table or generated by a calculator, a computer, or the Random Number Generator applet on the CD-ROM.

The students could decide to conduct the two-group experiment as a *blind study*. In such a case, the study participants would not know which group received which treatment (type of chips). Table 4.1 gives the results of a two-group experiment (which was not a blind study), comparing melting times of chocolate-flavored chips and semisweet chocolate chips.

Table 4.1.
Results of a Study (Using a Two-Group Design) Comparing the Time (in Seconds) to Melt Chocolate-Flavored Chips and Semisweet Chocolate Chips

Chocolate-Flavored Chips		Semisweet Chocolate Chips	
Student	Time to Melt (Seconds)	Student	Time to Melt (Seconds)
1	95	9	80
2	112	10	72
3	100	11	87
4	74	12	49
5	83	13	63
6	126	14	63
7	140	15	54
8	116	16	110

The students will use graphs to compare the two groups in their experiment. Parallel histograms, back-to-back stem-and-leaf plots, and parallel box plots are all good methods for them to use to study their data. The parallel box plots shown in figure 4.2 depict the data from table 4.1.

Summary measures, such as the mean, median, and range, can be calculated from these data, as can the standard deviation. For example, the sample median time required to melt the chocolate-flavored chips is 106 seconds, 38.5 seconds more than the sample median of 67.5 seconds

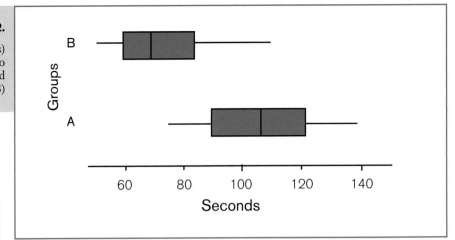

needed for the semisweet chocolate chips. Similarly, the sample mean times used to melt the chocolate-flavored and semisweet chocolate chips are 106 and 72 seconds, respectively, a difference of 34 seconds. However, the measures of spread for the two groups are remarkably similar. The sample ranges for the times needed to melt the chocolate-flavored chips and the semisweet chocolate chips are 66 and 61 seconds, respectively. The sample standard deviation is 22 seconds for the chocolate-flavored chips and 20 seconds for the semisweet chocolate chips. Although the two box plots show an overlap in the times needed to melt the two types of chips, the sample mean and median times needed to melt the chocolate-flavored chips are enough greater than those needed to melt the semisweet chocolate chips, relative to the spread of the data, for us to believe that it is unlikely that these differences are due to the chance variation that is present when the population mean difference is 0.

When looking at their data, students should think about possible sources of variation in their study. They may need to be reminded that they controlled some of the variation when designing the experiment. Yet, as in the data shown here, not all times were the same for a particular type of chip. The variation was possibly due to measurement error in the times it took the chips to melt, differences among chips (such as size and shape), and differences among people's mouths (such as temperature and chemistry of saliva). There could also be some variability that we cannot explain, such as different melting times for the same person with the same type of chip in repeated testing. Of these, the largest source of variation is usually the differences among people. A paired design can be used to account for some of these differences.

In the paired experiment, each student-subject melts one chip of each type in his or her mouth. The order in which the chip types are tested is randomly determined; that is, each student is randomly assigned to either a real semisweet chocolate or a chocolate-flavored chip as the first treatment. After measuring the time it takes to melt the first type of chip in the mouth, the student then receives the second type of chip as a treatment.

Sometimes students want to do just one randomization and use it to determine the order of treatments for all study subjects. For example, they might decide to flip a fair coin. Heads might mean that all subjects would melt the chocolate-flavored chip first and the real semisweet chocolate chip second, or vice versa. If students used this type of randomization and there turned out to be an advantage to being the first or the second treatment, then the observed differences in the times could be due to the order of the treatments and not to differences in the treatments themselves. Thus, it is important for each subject to have a separate randomization of the order in which he or she melts the real semisweet chocolate and the chocolate-flavored chips.

Students may decide that it is important for the subjects in the paired study to be blind to which type of chip they are testing. They may also decide that a certain time should pass between the two treatments or that subjects should drink a glass of water at a particular temperature between the treatments to "wash out" the effect of the first treatment.

After measuring the time for each type of chip to melt, students should take the difference in the two measurements in a consistent manner. That is, the difference could be the time for the real semisweet chocolate chip minus the time for the chocolate-flavored chip, or it could be the time for the chocolate-flavored chip minus that for the real semisweet chocolate chip. It should not be one type of difference for some participants and the opposite for others. Consistency in computation is necessary for positive and negative differences to have consistent meaning throughout the experiment. Once students have obtained the differences, they will use graphs and summary statistics to understand their data. Because person-to-person variability can be accounted for when a single student is tested with each chip in a paired design, this study should be easier for students to use to determine whether, on average, real semisweet chocolate or chocolate-flavored chips melt in the mouth more rapidly. Table 4.2 provides the results of a study with a paired design.

Table 4.2.
Results of a Study (Using a Paired Design) Comparing the Time (in Seconds) to Melt Chocolate-Flavored Chips and Semisweet Chocolate Chips

Student (Number = i)	Time in Seconds (x_i) for Student i to Melt a Chocolate-Flavored Chip	Time in Seconds (y_i) for Student i to Melt a Semisweet Chip	Difference in Times for Student i $d_i = x_i - y_i$
1	75	57	18
2	91	58	33
3	103	79	24
4	78	53	25
5	94	55	39
6	158	92	66
7	192	108	84
8	110	90	20

Because a paired design can generate twice as much data as a two-group design for a given number of study subjects, students often think that a paired design also yields twice as much information. However, if pairing is not an effective means of collecting data, researchers can actually lose information by using a paired design. A disadvantage of pairing by subject is that it may take twice as long to conduct the study. Because the time required to melt a chocolate chip in the mouth is not great, this disadvantage is not serious in the students' experiment, but for some studies, the time required to conduct the study is an important issue.

If the differences among people were not responsible for a significant portion of the variability in the measured times for melting the chocolate chips, then pairing would be of no value. For example, suppose there were not enough time for each participant to melt both kinds of chips, so the students decided to block participants on the basis of height, forming pairs from the shortest participants, the next shortest, and so forth, and randomly assigning one member of each pair randomly to one type of chip and the other to the other type. Height would not be an acceptable blocking criterion, because we would not expect it to have an effect on the time required to melt a chocolate chip.

In conducting the paired experiment with the chocolate chips, students may notice that if an individual subject takes longer than most others to melt a real semisweet chocolate chip, then he or she is also likely to take longer to melt a chocolate-flavored chip. One way to determine if this is the case is to plot the length of time for each subject to melt the chocolate-flavored chip against his or her time to melt the real semisweet chocolate chip. Figure 4.3 shows such a plot for the data in table 4.2. In the scatterplot shown, low time values for melting chocolate-flavored chips tend to be associated with low time values for melting semisweet chocolate chips, and likewise, high time values for melting chocolate-flavored chips tend to be associated with high time values for melting semisweet chocolate chips.

In cases where such an association holds true, each single subject can act as a block (that is, a pair) in a paired experiment. In the students' experiment with chocolate chips, blocking will enable them to account for the differences in people, thus permitting them to measure the differences in chocolate chip types more precisely. The box plot in figure 4.4 shows the differences between each subject's times to melt the two types of chips in the paired design.

As with the two-group design, students estimate summary measures for the differences. On average, it took an estimated 39 seconds longer to melt a chocolate-flavored chip than it did to melt a semisweet chocolate chip. The median additional time required to melt a

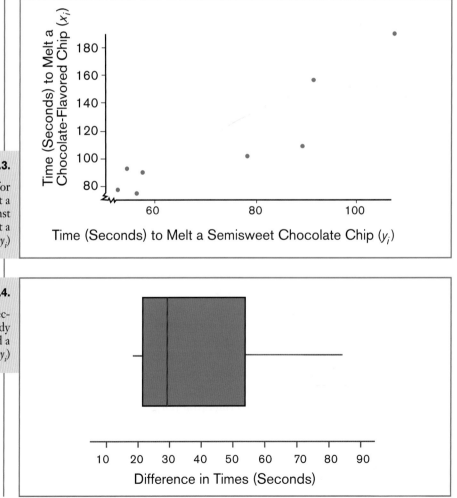

Fig. **4.3.**

A scatterplot of the time (in seconds) for each student i in the paired study to melt a chocolate-flavored chip (x_i) against his or her time to melt a semisweet chocolate chip (y_i)

Fig. **4.4.**

A box plot of the difference in time (in seconds) for each student i in the paired study to melt a chocolate-flavored chip (x_i) and a semisweet chocolate chip (y_i) ($d_i = x_i - y_i$)

chocolate-flavored chip was an estimated 29 seconds. The sample range of the additional time needed to melt a chocolate-flavored chip was 66 seconds, and the sample standard deviation of the differences in dissolving times for chocolate-flavored and semisweet chocolate chips was 24 seconds. The students should notice an overlap of the two groups when comparing times needed to melt the chips from the two-group experiment (refer to fig. 4.2), but they should observe that each person participating in the paired experiment took more time to melt the chocolate-flavored chip than the semisweet chocolate chip—that is, all differences were positive. The students' conclusion from each experiment should be that, on average, it takes longer to melt chocolate-flavored chips than semisweet chocolate chips. However, that conclusion is supported more strongly by the paired experiment than the two-group experiment because the paired design has accounted for person-to-person variability.

Possible Errors

Students should be encouraged to think about the hypotheses being considered here. The null hypothesis is that there is no difference in the average times required to melt chocolate-flavored chips and semisweet chocolate chips. The alternative hypothesis is that there is a difference, on average, in these times. Notice that the word *average* appears in the statement of both hypotheses. Although sometimes the chocolate-flavored chip may take less time to melt than the semisweet chocolate chip, and vice versa, the goal is to determine what happens on average.

Students should have found it fairly easy to conclude that a difference exists in the mean times required to melt the two types of chips. As the difference in the treatment means becomes smaller, it is not as simple to determine whether the observed difference is real or could be due to chance. Similarly, as the variability in the times to melt the chips increases, the box plots for the two-group experiment may overlap more or the one for the paired experiment may extend quite a bit to the left and right of zero, again making it difficult to determine whether the observed mean difference is real or a consequence of chance variation present when the population mean difference is 0. Formal statistical inference provides tools for drawing conclusions in these cases.

Conclusion

The results from research studies are reported daily on television, on radio, and in newspapers. In addition to the relationship between cellular phone use and cancer, studies have explored the effect of drivers' use of cellular phones on traffic accident rates. The effect that listening to different types of music while studying may have on schoolchildren's learning is another area of research interest. The efficacy of particular drugs and dietary supplements—for example, nonprescription diet drugs for weight loss and vitamin C to minimize colds—is also an active area of research that can have an impact on our lives. The challenge that often comes with these and other study topics is that different studies addressing what at least appears to be the same research question frequently report conflicting results. Encourage your students to

It is important for students to realize that rejecting a true null hypothesis results in an error of type I—not type II. Thus, if the students decided that there was a difference in the mean times required to melt the two types of chips, their only possible error would be one of type I, which they would commit if there was in fact no difference in the mean times. To make an error of type II, the students would need to conclude that there was no difference in the mean time needed to melt the two types of chips when in fact there was such a difference.

use the concepts of design to evaluate these reports critically and thus become better consumers of the volumes of information being generated in today's world.

Begin a collection of "research studies"

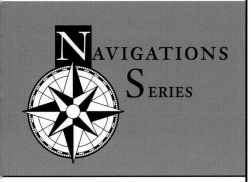

NAVIGATING *through* DATA ANALYSIS

Chapter 5
Putting It All Together

This chapter presents five problems that deal with the kinds of statistical thinking considered in the first four chapters. Each of the problems focuses on making decisions in the presence of uncertainty and involves elements such as carefully thought-out questions, a study design that accounts for major sources of variation, random selection of units or random assignment to treatments, and probability as a tool for determining whether an observed statistic could have occurred by chance variation under a specified null hypothesis. Teachers might use the problems to assess students' understanding of the big ideas that have been covered in this book:

- Sampling and the role of randomness
- Sampling distributions, summary statistics, and graphs to make sound decisions based on statistical thinking
- The design of studies

Problem 1: The Salk Polio Vaccine

p. 123–24

In 1954, the National Foundation for Infantile Paralysis (NFIP) and the U.S. Public Health Service were both preparing to conduct tests on the public at large of a vaccine developed by Jonas Salk. Infantile paralysis, also called poliomyelitis, is more commonly known as polio. The two groups approached the testing in different ways. Recognizing that children in grades 1–3 were most susceptible to polio, the NFIP focused its test on children in those three grades. The NFIP study used children in grade 2 as the treatment group and gave them the vaccine. Children in grades 1 and 3 were used as the control group. All the children had to have parental consent to be vaccinated, however, and many of the second graders' parents did not give their consent (see table 5.1; note that the numbers are rounded).

Table 5.1
Polio Vaccine—NFIP Study

	The NFIP Study	
	Size	Rate of Polio Cases per 100,000
Grade 2 (vaccine)	225,000	25
Grades 1 and 3 (control)	725,000	54
Grade 2 (no consent)	125,000	44

(Francis 1955)

The Public Health Service researchers used a different design. They selected a set of students whose parents had given consent and randomly assigned those children to the treatment (vaccine) or to the control group. All children who did not get the vaccine were given a placebo (a harmless injection of salt and water), so the children did not know whether they were given the treatment or not. In addition, the doctors who subsequently treated children with suspected polio did not know whether the children had received the vaccine or a placebo. Such an experiment is called *double-blind* because neither the subjects nor the evaluators know who has had the treatment. The results of this study appear in table 5.2 (note that the numbers are rounded).

Table 5.2
Polio Vaccine—Public Health Study

	Randomized Controlled Double-Blind Experiment	
	Size	Rate of Polio Cases per 100,000
Treatment	200,000	28
Control	200,000	71
No consent	350,000	46

(Francis 1955)

In considering the two studies, students are asked to do the following:

- Comment on the two methods with respect to the question of interest, the design of the studies, the role of randomization, and the results.
- Show how two-way tables can help in analyzing the problem.
- Contrast the populations being sampled in the two cases. Could the differences between the populations make a difference in the outcomes of the two studies?

Discussion

In both studies with the Salk vaccine, the question of interest was whether the percentage of vaccinated children who subsequently contracted polio was significantly smaller than the percentage of unvaccinated children who contracted the disease. In several respects, the

"All students should ... know the characteristics of well-designed studies, including the role of randomization in surveys and experiments."
(NCTM 2000, p. 324)

Navigating through Data Analysis in Grades 9–12

NFIP study was poorly designed. For one thing, the control and treatment groups were not comparable. The treatment group consisted of second-grade students whose parents had consented to the vaccine. The control group of first and third graders presumably consisted of both children whose parents would have consented and children whose parents would not have consented. Consent was not considered, since no first or third grader received the treatment.

Researchers realized that the NFIP study was flawed, though not for reasons that typical readers would immediately see. The study was biased against the vaccine, because the grade 2 children who received the treatment were more likely to get polio than those who did not. The reason is essentially as follows: Grade 2 students whose parents consented were being compared with all the students from grades 1 and 3, whether those students' parents would have consented or not. The parents who allowed their children to be vaccinated were more likely to have higher incomes than those who did not. It was also true that children who lived in less hygienic conditions tended to contract mild cases of polio while they were very young and were still protected by antibodies from their mothers. Once such children contracted polio, their systems generated antibodies that prevented them from getting more severe cases of the disease as they grew older. The children of the wealthier, "consenting" parents typically lived in more hygienic conditions, had not contracted polio in any form while they were young, and as a result did not have the protective antibodies. Thus, when these children became older, they were more likely to contract polio.

A second possible concern about the design of the NFIP study involved the fact that polio is a contagious disease and spreads through contact. It could have been the case that the incidence of polio for children in grade 2 was higher than that for children in grades 1 and 3, biasing the study against the vaccine. Or, by the same token, the incidence might have been lower, biasing the study in favor of the vaccine. The design did not take into account or control for the possible effects of the spreading of the disease (Freedman, Pisani, and Purves 1998).

In contrast, the study conducted by the Public Health Service took as its population children whose parents gave consent. As discussed earlier, these parents tended to differ from those who did not give consent. What impact did this difference have on the study? Randomness helped ensure that the two groups were similar with respect to the important variables, thus reducing the chance of biased results. A sample of children from this population was randomly divided into two groups, one receiving the vaccine and the other receiving a placebo, with neither the children nor their parents knowing which children were receiving the vaccine. Because many types of polio are difficult to diagnose, the researchers were concerned that a doctor's diagnosis in an ambiguous case might be affected by knowing whether the child had received the vaccine or not. Thus, the double-blind design made sure that even the doctors who treated the children who were suspected of having contracted polio did not know who had received the vaccine.

Basically, the situation in this study is similar to that in the scenario Discrimination or Not? in chapter 2. If the vaccine had no effect, researchers would have expected about the same incidence of polio in the two groups (treatment and control), since the groups were of equal size. That is, of the 198 children who contracted polio, about half

Randomization helped ensure that the two groups were similar with respect to the important variables, thus reducing the chance of biased results.

would be in each group. Researchers would have expected about 99 cases in the vaccinated group and about 99 in the control group, but instead they found 56 cases among vaccinated children and 142 among those who did not receive the vaccine (see tables 5.3 and 5.4). Was this difference due to chance variation or could it have been due to some other factor, such as the vaccine?

Simulating the study would be unwieldy because of the large numbers. Formal statistical techniques, based on reasoning about sampling distributions, can be used to answer the question. Two-way tables can be used to organize the data from the study and to illustrate the expected outcomes if the tests clearly showed no difference in the groups as a result of the vaccine.

Table 5.3
Polio Vaccine—Public Health Study Results

	Contracted Polio	Did Not Contract Polio	Total
Treatment	56	199,944	200,000
Control	142	199,858	200,000
Total	198	399,802	400,000

Table 5.4
Expected Results for the Public Health Study If the Treatment (Vaccine) Had No Effect

	Contracted Polio	Did Not Contract Polio	Total
Treatment	99	199,901	200,000
Control	99	199,901	200,000
Total	198	399,802	400,000

Problem 2: Medical Studies

p. 125–26

Study 1: Sleep apnea may be linked to Alzheimer's

Hiroshi Kadotani and others (2001) reported in the *Journal of the American Medical Association* that "a disorder that causes breathing lapses during sleep is linked to a gene variation associated with Alzheimer's and cardiovascular problems." The researchers wrote that sleep-disordered breathing could be a result of a gene variation called apolipoprotein E-4. Sleep apnea, characterized by snoring and brief breathing lapses during sleep, is mostly hereditary. In the study, the risk of sleep apnea for participants with the E-4 gene variation was almost double the risk of sleep apnea for those without the trait. The sleep disorders study involved 791 middle-aged adults, ranging from 32 to 68 years of age. Of the 222 participants with the E-4 variation, 27 had moderate to severe apnea, as compared with 39 of the 569 participants without the variation. According to the study, a significant portion of sleep-disordered breathing is associated with the gene variation in the general population. The authors said that sleep apnea may interact with the E-4 trait to impair mental abilities, but their study does not necessarily imply that all adults with sleep apnea face an increased risk of Alzheimer's or cardiovascular disease.

Study 2: Tirofiban improves the success rate of bypass surgery

Gilles Montalescot and others (2001) found that the heart drug tirofiban helped patients who were undergoing bypass surgery. This study randomly assigned 300 patients in a double-blind fashion to either tirofiban (149 patients) or a placebo (151 patients) before they underwent coronary angiography, a procedure to expand the artery. After thirty days, death or major setbacks had occurred in 9 of those in the tirofiban group as compared with 32 of those in the placebo group. The investigators concluded that early administration of tirofiban improved the success rate of the bypass surgery.

Students are asked to discuss the summaries of studies 1 and 2 with respect to the following points:

- Were the investigations observational studies or experiments?
- What were the questions of interest?
- How were the studies designed?
- What role did randomization play in the studies?
- What conclusions could be drawn from the studies?

Discussion

Study 1

The report by Kadotani et al. (2001) describes an observational study with no random assignment of treatments (gene variation) or random selection of study participants from the population of middle-aged adults. The subjects were categorized as those with the gene variation or those without the gene variation. The response variable was moderate to severe sleep apnea, and the observational unit was a study participant. Table 5.5 summarizes the numbers for participants in the study.

Table 5.5
Study Results on the Gene Variation Apolipoprotein E-4 and Sleep Apnea

	Moderate to Severe Sleep Apnea	No Significant Sleep Apnea	Total
Gene variation	27	195	222
No gene variation	39	530	569
Total	66	725	791

To answer the question of whether the data support the claim that sleep apnea is related to the gene variation, students should compute the numbers of sleep apnea cases that they would expect in each group of participants—those with and those without the gene variation—if no relationship existed. The fraction of all study participants with sleep apnea was 66/791, or 8 percent. If no relationship existed between sleep apnea and the gene variation, students would expect participants with the gene variation and those without it to exhibit sleep apnea in about the same proportion—the ratio of 66/791 that was observed for all study participants. Table 5.6 shows the computation of the expected counts if there were no relationship between sleep apnea and gene variation.

Note that in table 5.6 the expected number of sleep apnea cases is greater for the "no gene variation" group than for the "gene variation" group, but the former group is larger than the latter. It is the proportion of cases—not the number of cases—that students would expect to be the same for both groups if no relationship existed between sleep apnea and the gene variation. Students should also realize that after

"All students should ... develop and evaluate inferences and predictions that are based on data." (NCTM 2000, p. 324)

Table 5.6

Expected Results If No Relationship Existed between the Gene Variation Apolipoprotein E-4 and Sleep Apnea

	Sleep Apnea	No Sleep Apnea	Total
Gene variation	$\dfrac{66}{791} \cdot 222 \approx 19$	$\dfrac{725}{791} \cdot 222 \approx 203$	222
No gene variation	$\dfrac{66}{791} \cdot 569 \approx 47$	$\dfrac{725}{791} \cdot 569 \approx 522$	569
Total	66	725	791

computing one of the expected counts, they can find the others by subtracting, using the row or column totals.

The question of interest then becomes "Are the differences between the observed counts (in table 5.5) and the expected counts (in table 5.6) due to chance variation, or are they the result of a relationship between sleep apnea and the gene variation apolipoprotein E-4?" To answer this question, students could use a simulation, as they did in chapter 2, to examine how likely it would be for 27 or more participants with both sleep apnea and the gene variation to occur in such a study by chance variation if no relationship existed between sleep apnea and the gene variation. For example, students could take the numbers 1–791 and let 1–222 represent the study participants with the gene variation. From the numbers 1–791, they could then generate 66 random numbers without replacement to represent the study participants with moderate to severe sleep apnea. (There should be no repetition of random numbers, because patients should not be counted twice.) Then they could count those numbers that were less than 222 to obtain one statistic in the simulated sampling distribution. If they did this a sufficient number of times, then they would be likely to conclude, as the researchers in this study did, that the differences in the observed and expected counts were extremely unlikely to be due solely to chance variation and that there was a strong association between the gene variation and sleep apnea.

Although the researchers in the study concluded that there was a strong association between the gene variation and sleep apnea, they could not identify a causal relationship, because their study was observational and not an experiment. The coincidence of the two factors, sleep apnea and gene variation, could have been due to other conditions not accounted for in the study.

Study 2

In the study by Montalescot et al. (2001), the researchers randomly assigned 300 patients to either a treatment or a control group in order to evaluate the effects of the drug tirofiban before bypass surgery. Table 5.7 summarizes the results.

Table 5.7

Results (after 30 Days) of the Study of Tirofiban and Bypass Surgery

	Major Setbacks/Deaths	No Major Setbacks	Total
Tirofiban	9	140	149
Placebo	22	129	151
Total	31	269	300

Because the patients are randomly assigned to treatments, the study is an experiment. Patients appear to have been recruited instead of randomly selected, and this technique would limit the scope of inference to those individuals in the study.

Suppose that no relationship existed between tirofiban and outcomes of bypass surgery. How many patients within each treatment group would students expect to have a major setback or to die? Because the numbers in the two groups are about the same, students should understand that about half of the 31 instances of major setbacks/deaths should occur in each group. Thus, they might expect around 15 or 16 of the patients who took tirofiban to have suffered major setbacks or to have died, as shown in table 5.8. (Here 15 is used, but 16 would be equally correct.)

Table 5.8
Results Expected (after 30 Days) If No Relationship Existed between Tirofiban and Bypass Surgery Outcomes

	Major Setbacks/Death	No Major Setbacks	Total
Tirofiban	15	134	149
Placebo	16	135	151
Total	31	269	300

The statistical question of interest is whether the observed number of bypass patients who received tirofiban and who also suffered major setbacks or died—9—was smaller than the expected number—15—by a difference that was only the result of chance variation with no relationship existing between tirofiban and improved outcomes of bypass surgery. In other words, how much variability would students expect to find around 15 in this study if tirofiban had no more effectiveness than a placebo?

Students could use a simulation to generate a simulated sampling distribution, as in chapters 2 and 3. They could assemble 300 cards, with 149 red and 151 black cards representing, respectively, the bypass patients who received tirofiban and those who did not receive the drug. Students could draw 31 cards to represent all the patients who subsequently died or suffered major setbacks, and then they could record the number of red cards representing those who were taking tirofiban. Simulated sampling distributions would typically support the claim that 9 was an unusual outcome under the null hypothesis and, thus, that the reduction in the number of patients with adverse effects was probably due to a cause other than chance variation—in this case, the drug tirofiban. Students should see that consequently it would be reasonable to assert that use of tirofiban had a significant and positive effect on bypass surgery outcomes.

Problem 3: Darwin and Genetics

p. 127

Charles Darwin was interested in whether there was any difference in the height of plants that were cross-fertilized—that is, that were fertilized by pollen from other, closely related plants—and those that were self-fertilized. The data in table 5.9 show the height in inches of plants grown in pairs for a fixed period of time. One member of each pair was cross-fertilized, and the other was self-fertilized.

Table 5.9.
Plant Height

Pair	Cross-Fertilized (in.)	Self-Fertilized (in.)
1	23.5	17.4
2	12.0	20.4
3	21.0	20.0
4	22.0	20.0
5	19.1	18.4
6	21.5	18.6
7	22.1	18.6
8	20.4	15.3
9	18.3	16.5
10	21.6	18.0
11	23.3	16.3
12	21.0	18.0
13	22.1	12.8
14	23.0	15.5
15	12.0	18.0

(Darwin 1876, p. 451)

Students are asked to examine the data and decide what conclusions they can draw from them. They are asked to justify their conclusions by using techniques discussed in class from this book.

Discussion

Little information is given about how the plants were paired or how randomization was used to assign the treatment. Assuming that the study was well designed in this respect, students can compare the differences much as they compared differences in chapter 3. A scatterplot showing the line *Cross-fertilization = Self-fertilization* as in figure 5.1 can demonstrate that in all but two pairs, the cross-fertilized plant is larger than the self-fertilized one. Students should note that in this case, a scatterplot with the axes reversed would be equally informative. In this scatterplot, all but two dots would be above *y = x*.

A dot plot that shows each pair by the height of the cross-fertilized plant minus the height of the self-fertilized plant can also demonstrate

"All students should … select and use appropriate statistical methods to analyze data … [and] develop and evaluate inferences and predictions that are based on data."

(NCTM 2000, p. 324)

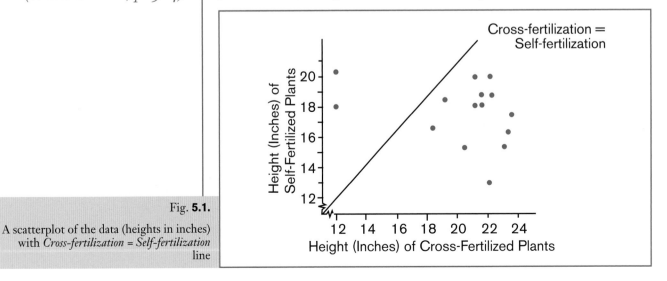

Fig. 5.1.

A scatterplot of the data (heights in inches) with *Cross-fertilization = Self-fertilization* line

Navigating through Data Analysis in Grades 9–12

that, with two exceptions, the differences are positive. Such a plot is shown in figure 5.2.

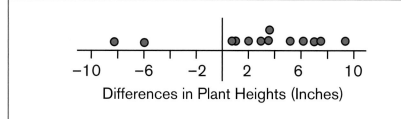
Differences in Plant Heights (Inches)

Fig. **5.2.**

A dot plot showing *Cross-fertilized* minus *Self-fertilized*

If there were no difference between the two methods of fertilization in their effects on the heights of plants, students could expect the sample mean difference to be zero. However, the sample mean difference is 2.6. How likely would a sample mean of 2.6 be to occur by chance variation if the population mean differences were zero?

By simulating the situation as they did in chapter 3, students can create a sampling distribution to help answer the question. Figure 5.3 shows sample means of the random differences from 100 simulations. The simulated probability that the observed outcome—a mean difference of 2.6—occurred by chance is just 2 percent if no difference existed in the mean heights of self-fertilized and cross-fertilized plants. This would seem to be strong evidence that the difference between the heights of self-fertilized plants and cross-fertilized plants is greater than can be explained by chance variation.

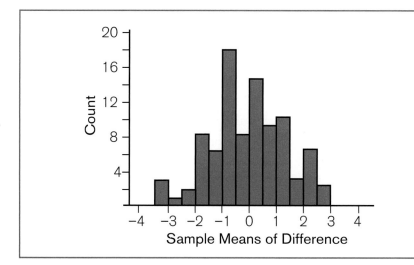

Fig. **5.3.**

A simulated sampling distribution of sample means of random differences from 100 simulations

"All students should … understand and apply basic concepts of probability." (NCTM 2000, p. 324)

Problem 4: Examining an Article in the News Media

This problem asks students to take the next step: they are to consult the news media to find an article describing a study. Referring to the report, they are asked to write a description of the study, being sure to include the following components in their description:

• The question under investigation

p. 128

- The design of the study
- The nature of the data collected
- Any assumptions made by the researchers
- The role of randomization
- The results
- An interpretation of the conclusions

Discussion

Making a two-way table to organize the information can often help students make sense of articles that appear in the media. In some cases, students can trace the studies mentioned in these articles back to their sources by a search of the Web. If they can locate the journal in which the report of the study originally appeared, the abstract and other sections of the report may be helpful and interesting as resources. Students' descriptions of the study should include the question of interest, how the study was conducted, who sponsored the work, information about the design of the study, and a statement about the findings.

Problem 5: Designing a Study

Instead of asking students to consider a completed study—either one presented to them or one that they have located—this problem calls on them to design a study of their own to explore a question of particular interest to them. They are asked to state the question, any assumptions, data that are important and how they will collect them, the role of randomization, and how they propose to use probability in drawing conclusions. Students might investigate questions such as the following:

- Does studying with the television on make a difference in students' grade-point averages?
- Does regular exercise lower pulse rate?
- Is there a difference between two brands of microwave popcorn in the amounts popped?
- What characteristics of a paper airplane enable it to fly farther?

Discussion

Students may need some guidance in designing a study to answer questions such as those listed. After they have selected a problem, you might ask them to consider the following preliminary questions:

- What do I want to know? What question do I want answered?
- What data will help me find an answer? (You can either have students imagine that they have a set of data or actually have them do a small pilot test to collect some. Then the students can be encouraged to consider how they will organize and analyze their data to check whether the data they intend to collect will in fact be useful.)
- What considerations should I make before I start collecting the data? Will I collect the data from an experiment, a sample survey, or an observational study? What do I need to do to be sure that the study is well designed? Are my questions carefully worded?

p. 129

"All students should ... formulate questions that can be addressed with data and collect, organize, and display relevant data to answer them."

(NCTM 2000, p. 324)

- What is the parameter of interest for the population?
- What would be an appropriate statistic for me to use to estimate the parameter?
- When I display my data, what methods can I use to help answer the question of interest?

By actually designing their own studies, students have the opportunity to learn from their mistakes.

Conclusion

Many of the ideas illustrated in the preceding chapters will begin to make sense as students work through the process of designing studies on their own. For example, they have to resolve how to choose a sample, and, depending on their question, consider comparable packages of popcorn or determine how to launch paper airplanes in a consistent manner. One resource is Exploring Projects (Errthum et al. 1999).

NAVIGATIONS SERIES

GRADES 9–12

NAVIGATING *through* DATA ANALYSIS

Looking Back and Looking Ahead

Reasoning from data to make decisions is a critical skill for students entering the world in which they will work and live as adults:

> In grades 9-12 students should gain a deep understanding of the issues entailed in drawing conclusions in light of variability. They will learn more-sophisticated ways to collect and analyze data and draw conclusions from data in order to answer questions or make informed decisions in workplace and everyday situations. They should learn to ask questions that will help them evaluate the quality of surveys, observational studies, and controlled experiments. (NCTM 2000, p. 325)

The activities in *Navigating through Data Analysis in Grades 9–12* are intended to give students this experience by providing a foundation for thinking about and working with data in interesting and relevant contexts.

NAVIGATIONS
SERIES

GRADES 9–12

NAVIGATING
through
DATA ANALYSIS

Appendix
Blackline Masters and Solutions

Random Rectangles

You will explore this population of 100 rectangles in the following activities:

- Sampling Rectangles
- Sample Size
- Sampling Methods

Navigating through Data Analysis in Grades 9–12

Sampling Rectangles

Name _____

Reports of polls often include the information that the sample populations for the polls were "randomly selected." Why do polls use data from people selected randomly? Can't most people simply do a good job of choosing a "typical" group of people for a poll? Why do scientists choose random samples when they are designing experiments, and how do they make their selections?

For this activity, your teacher will give you a copy of the activity sheet "Random Rectangles," which shows rectangles of different areas. Keep the sheet covered until your teacher gives the signal to begin!

1. Look at the rectangles on the activity sheet. Each small square represents a rectangle that has an area of one. Select five rectangles that you think would give a good representation (that is, a representative or "typical" sample) of all of the rectangles on the sheet.
   a. Record the numbers of the rectangles that you chose and give their corresponding areas.
   b. Compute and record the sample mean area for the five rectangles that you selected.

"Typical" Rectangles

Number of rectangle   Area

1._____   _____

2._____   _____

3._____   _____

4._____   _____

5._____   _____

Sample mean area of the "typical" rectangles _____

2. Generate five random numbers between 1 and 100 by using a random number generator on a calculator or computer. If a number repeats, discard it, and generate another one to replace it.
   a. Use these random numbers to locate the five rectangles that have the corresponding numbers on the sheet, and record their areas.
   b. Compute and record the sample mean area of the five rectangles "selected" by the random numbers.

Random Rectangles

Number of rectangle   Area

1._____   _____

2._____   _____

3._____   _____

4._____   _____

5._____   _____

Mean area of the random rectangles _____

Sampling Rectangles (continued)

Name _____

3. Report to the class the sample mean for the areas of the set of rectangles that you chose as a "typical" sample and the sample mean for areas of the the set of rectangles that you chose randomly. Compile the class results and work with your classmates to make dot plots of the two simulated *sampling distributions*—one for the sample means of the subjectively chosen samples of rectangles and one for the sample means of the randomly chosen samples of rectangles. Compare the simulated sampling distribution of the sample mean areas that you found from random sampling with the simulated sampling distribution that you found from your subjectively chosen samples. What can you observe?

4. Using the activity sheet "Data Record," record a measure of center and a measure of spread (median and interquartile range, or mean and standard deviation) for the simulated sampling distribution of the random samples of size 5. Write a brief description, including shape, center, and spread, of the simulated sampling distribution.

Distribution of Sample Mean Areas of Rectangles

Name _____

Sample Mean Areas of Subjectively Chosen Rectangles

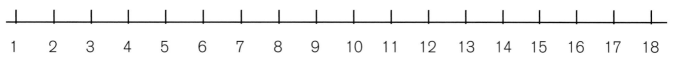

Sample Mean Areas

Sample Mean Areas of Randomly Chosen Rectangles

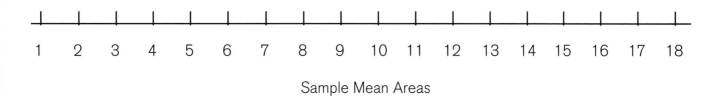

Sample Mean Areas

Sample Size

Name _____

Follow your teacher's directions for this activity.

1. Using the 100 rectangles pictured on the activity sheet "Random Rectangles" and some type of random number generator, randomly select 10 rectangles and calculate their mean area.

2. Collect everyone's results and make a dot plot of the simulated sampling distribution of the sample means for the samples of size 10.

   a. How does this new simulated sampling distribution compare with the simulated sampling distribution that your class made in the activity Sampling Rectangles for the means of the random samples of size 5?

   b. In particular, how does the sample size affect the shape, center, and spread of the two simulated sampling distributions?

3. Using the activity sheet "Data Record," enter a measure of center and a measure of spread (median and interquartile range, or mean and standard deviation) for each of the two simulated sampling distributions, and write a brief description of the simulated sampling distribution.

Data Record

Name _____

Make a rough sketch of your simulated sampling distribution in each case. Record a measure of center and a measure of spread, or variability—either the mean and the standard deviation or the median and the interquartile range.

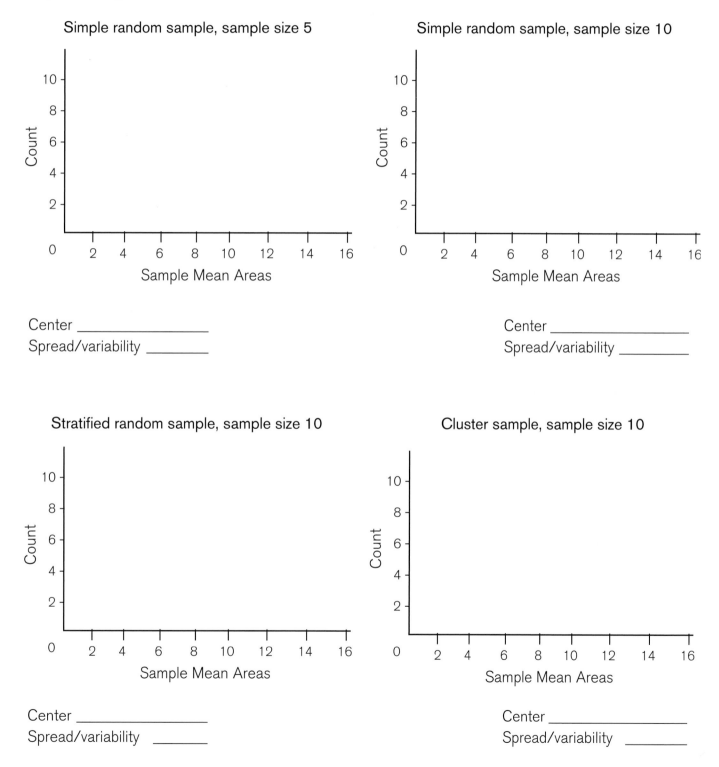

Simple random sample, sample size 5

Center _____

Spread/variability _____

Simple random sample, sample size 10

Center _____

Spread/variability _____

Stratified random sample, sample size 10

Center _____

Spread/variability _____

Cluster sample, sample size 10

Center _____

Spread/variability _____

Sampling Methods

Name _____

Task A—Stratified Samples

Suppose you were again using a sample to estimate the population mean of the areas of the rectangles shown on the activity sheet "Random Rectangles," and this time you wanted to make sure that your sample contained rectangles with both small and large widths.

Tables 1 and 2 show the set of 100 rectangles divided into two groups (strata) according to the widths of the rectangles. Note that in this division of the rectangles, *width* indicates a rectangle's horizontal dimension. One stratum contains 59 rectangles with widths less than 3 (table 1), and the other stratum contains 41 rectangles with widths greater than or equal to 3 (table 2).

Table 1

Stratum of Rectangles with Widths Less than 3, Listed by the Numbers Corresponding to Them on "Random Rectangles"

1	2	3	4	5	6	8	9	10	11	12	14	16	18	19
20	21	22	24	27	28	30	32	34	35	36	40	43	44	45
46	48	49	50	51	53	56	62	64	67	68	69	71	73	74
75	77	78	79	80	82	83	84	87	88	89	91	95	100	

Table 2.

Stratum of Rectangles with Widths Greater than or Equal to 3, Listed by the Numbers Corresponding to Them on "Random Rectangles"

7	13	15	17	23	25	26	29	31	33	37	38	39	41	42
47	52	54	55	57	58	59	60	61	63	65	66	70	72	76
81	85	86	90	92	93	94	96	97	98	99				

Stratified random sampling is a process that involves randomly selecting samples from groups like these, each of which is considered a stratum of the population.

Use the rectangles in tables 1 and 2 to explore the process and results of stratified random sampling:

1. Using some type of random number generator, randomly select 5 rectangles from each table (stratum), and then compute the mean area of each stratum of 5 rectangles. The sample mean for the combined strata is found by using the population proportion, as follows:

$$\frac{59}{100} \cdot (\text{mean of stratum from table 1}) + \frac{41}{100} \cdot (\text{mean of stratum from table 2})$$

Sampling Methods (continued)

Name _____

Repeat this process three times.

2. Compile the class data and show all sample means in a simulated sampling distribution. Describe the shape, center, and spread of the simulated sampling distribution.

3. Using the activity sheet "Data Record," sketch the simulated sampling distribution and enter a measure of center and a measure of spread (median and interquartile range, or mean and standard deviation) for the simulated sampling distribution of sample mean areas from stratified sampling. Write a brief description of this simulated sampling distribution.

Sampling Methods (continued)

Name _____

Task B—Cluster Samples

Sometimes a population of interest has so many members or the members are so dispersed that the cost of taking a simple random sample is too high. In such instances, researchers may use *cluster sampling,* a random selection process by which clusters of individuals are identified in the population, and the individuals in the clusters are then studied.

Table 3 shows the 100 rectangles from the activity sheet "Random Rectangles" divided into twenty clusters. Each cluster is "like" every other cluster in that it contains 5 rectangles that are relatively close together on the sheet. Each also contains as much variability as is possible in its particular "neighborhood." Each cluster is named by a Roman numeral.

Table 3.
Clusters of Rectangles Listed by the Numbers Corresponding to Them on "Random Rectangles"

I	II	III	IV	V	VI	VII	VIII	IX	X
1	3	8	7	19	18	25	40	32	39
2	4	13	11	26	20	34	41	33	47
9	5	14	12	27	21	35	42	38	48
16	6	15	22	28	30	36	43	45	49
17	10	24	23	29	31	37	44	46	50

XI	XII	XIII	XIV	XV	XVI	XVII	XVIII	XIX	XX
51	55	58	68	64	66	79	91	88	83
52	56	59	69	72	75	80	92	89	84
53	57	60	70	73	76	81	93	96	90
54	62	65	71	74	77	86	94	97	99
61	63	67	78	82	83	87	95	98	100

Use the clusters of rectangles in table 3 to explore the process and results of cluster sampling:

1. Using a random number generator, select two of the twenty clusters. Find the mean area of the your sample of 10 rectangles from the two clusters. Repeat the process three times.

Sampling Methods (continued)

Name _____

2. Compile the class data and show all sample means in a simulated sampling distribution.

3. Complete the activity sheet "Data Record," entering for this simulated sampling distribution a measure of center and a measure of spread (median and interquartile range, or mean and standard deviation). Write a description of the simulated sampling distribution, including center, spread, and shape.

4. Study the information that you have accumulated on the sheet "Data Record" and compare the sampling methods that you have studied.

   a. What can you say about the effect of sample size?

   b. What observations can you make about the three methods that you have explored: simple random sampling, stratified random sampling, and cluster sampling?

Discrimination or Not?

In 1972, 48 male bank supervisors were each given a personnel file and asked to judge whether the person represented in the file should be recommended for promotion to a branch-manager job described as "routine" or whether the person's file should be held and other applicants interviewed. The files were identical except that half of the supervisors had files labeled "male" while the other half had files labeled "female." Of the 48 files reviewed, 35 were recommended for promotion.

(From Rosen, Benson, and Thomas H. Jerdee, "Influence of Sex-Role Stereotypes on Personnel Decisions," *Journal of Applied Psychology* 59 [February 1974], pp. 9–14)

If you knew the numbers of "male" and "female" folders selected for promotion, and the selected "male" folders outnumbered the selected "female" folders, could you conclude that discrimination against women played a role in the bank supervisors' recommendations?

You will be exploring this *question of interest* in the following activities:

• What Would You Expect?

• Simulating the Case

• Analyzing Simulation Results

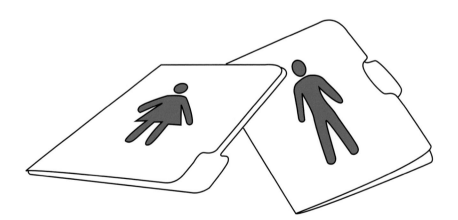

What Would You Expect?

Name _____

Read (or reread) the scenario presented in the activity sheet "Discrimination or Not?" The steps in the following activity will help you take a closer look at the data in the scenario. In answering the following questions, assume that 48 candidates were considered for promotion, as in the actual study.

1. Suppose that the recommendations of male and female candidates showed no evidence of discrimination on the basis of gender.

   a. How many males would you then expect to be recommended for promotion? How many females?

   b. Enter the values in table 1.

Table 1

No Discrimination by Gender

	Recommended for Promotion	Not Recommended for Promotion	Total
Male	_____	_____	24
Female	_____	_____	24
Total	35	13	48

2. Now suppose that the recommendations for promotion showed strong evidence of discrimination against the female candidates for promotion. Complete table 2 to show a possible example of this case.

Table 2

Strong Evidence of Discrimination against Women

	Recommended for Promotion	Not Recommended for Promotion	Total
Male	_____	_____	24
Female	_____	_____	24
Total	35	13	48

3. Suppose the evidence of discrimination against the women fell into a "gray" area, making any discrimination against the women not clearly obvious without further investigation. Complete table 3 to show such a case.

Table 3

"Gray" Evidence of Discrimination against Women

	Recommended for Promotion	Not Recommended for Promotion	Total
Male	_____	_____	24
Female	_____	_____	24
Total	35	13	48

What Would You Expect? (continued)

Name _____

4. In the actual situation depicted in the scenario, the results were that out of 24 files labeled "male," 21 candidates were recommended for promotion. Out of 24 files labeled "female," 14 candidates were recommended. Enter the data from the actual discrimination study in table 4.

Table 4
Actual Results of Discrimination Study

	Recommended for Promotion	Not Recommended for Promotion	Total
Male	_____	_____	_____
Female	_____	_____	_____
Total	_____	_____	_____

5. Consider the numbers of males and females recommended for promotion.

   a.   What percentage of the recommended candidates were male?
   b.   What percentage of the recommended candidates were female?

6. Without exploring the data any further, tell whether you think that the bank supervisors discriminated against the female candidates for promotion? How certain are you?

7. How likely do you think it is that chance variation was responsible for the smaller number of female applicants recommended for promotion? Explain your thinking.

Discussion and Extension

1. Suppose that a group of bank supervisors looked at the files of 48 actual applicants who were basically identical in their qualifications. Of the 35 applicants that the supervisors recommended for promotion, 21 were males, and 14 were females. If a lawyer retained by the female applicants hired you as a statistical consultant, how would you go about obtaining evidence to decide whether the observed results were due to chance variation or whether they could be due to discrimination against the women?

2. What thoughts do you have at this point about the manner in which the study was conducted? What would you need to assume about the study in order to infer that gender was the cause of the apparent difference between the number of females and the number of males recommended for promotion?

Simulating the Case

Name _____

Using a regular deck of 52 playing cards, remove 2 red cards and 2 black cards. Let the 24 remaining black cards represent the "male" files in the Discrimination or Not? scenario and the 24 remaining red cards represent the females. These 48 cards will simulate the 48 folders, half of which were labeled "male" and the other half, "female."

1. Shuffle the 48 cards at least six or seven times to ensure that any cards that you are going to count are from a random process.

2. Count out the top 35 cards. Let these cards represent the applicants recommended for promotion by the bank supervisors. (Alternatively, you could conduct the simulation more efficiently by dealing out 13 cards and letting them represent the applicants who were not recommended. This process would be equivalent to dealing out 35 "recommended" cards, and it would let you tally numbers of recommended males and females more quickly and easily.)

3. Out of the 35 cards, count the number of black cards (representing the recommended males).

4. Create a dot plot by placing a blackened circle or X above the number of black cards counted each time. The range of values for possible black cards is 11 to 24.

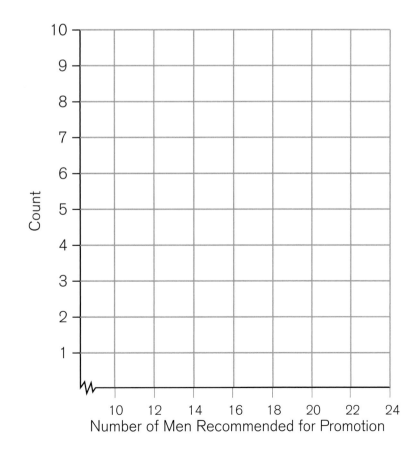

Name _____

5. Repeat steps 1–4 nineteen more times for a total of 20 simulations.

6. Using the results (the counts) plotted on the number line, estimate the chances that 21 or more black cards (males) out of 35 would be selected if the selection process were random; that is, if there were no discrimination against the women in the selection process.

7. Look at the dot plot, and comment on the shape, center, and variability of the simulated sampling distribution of the counts by answering the following questions:

   a. Is the simulated sampling distribution somewhat symmetrical, pulled (skewed) to the right, or pulled to the left?

   b. Do you observe any unusual counts?

   c. Where on the dot plot is the median of your observations?

   d. Estimate the mean of the simulated sampling distribution representing the number of males recommended for promotion from the 20 simulations.

   e. What counts occurred most often?

   f. Use the dot plot to comment on the spread of the data.

8. Is the shape of this simulated sampling distribution what you might have expected? Why, or why not?

Discussion and Extension

1. Think about the question of possible discrimination against the women. On the basis of your simulations, does there appear to be evidence to support a claim that recommending 21 males out of 35 candidates recommended for promotion was due to discrimination against the women rather than to chance variation? That is, how do your simulated results compare with those of the original study?

2. How do your simulated results compare with those of your classmates?

Analyzing Simulation Results

Figures 1–3 show plots of three different sets of 20 simulations.

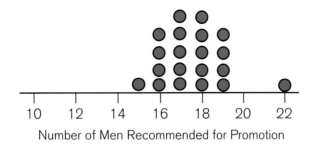

Fig. 1. Number of men recommended for promotion in 20 simulations

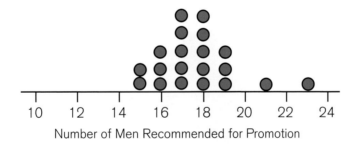

Fig. 2. Number of men recommended for promotion in 20 simulations

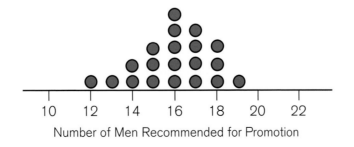

Fig. 3. Number of men recommended for promotion in 20 simulations

Analyzing Simulation Results (continued)

Name _____

Answer questions 1–4 for each of the three simulated sampling distributions that the plots show.

1. Describe the simulated sampling distribution of counts with respect to shape, center, and spread.

2. Is this simulated sampling distribution what you might have expected? Why, or why not?

3. Use the simulated sampling distribution to estimate the chance, or probability, that 21 or more of the 35 recommended for promotion would be male.

4. Do you believe that the simulated results provide evidence to support a claim that recommending 21 males for promotion was not due to chance variation and that there could be discrimination against women? Justify your answer.

Figure 4 shows a plot of the number of men recommended in 1291 simulations. Answer questions 5–7 on the basis of the data shown in the figure.

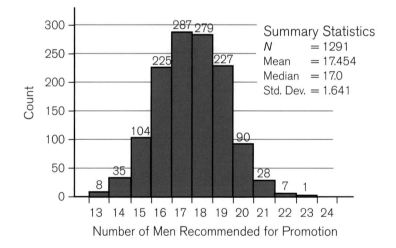

Fig. 4. Number of men recommended for promotion in 1291 simulations

5. Describe the simulated sampling distribution of counts with respect to shape, center, and variability.

6. Is this simulated sampling distribution what you might have expected? Why, or why not?

7. There are 36 counts of 21 or more in figure 4. What would you estimate as the chance, or probability, that 21 or more of the 35 selected for promotion would be male?

Analyzing Simulation Results (continued)

Name _____

8. On the basis of your explorations in questions 5–7, do you believe that there is evidence to support a claim that selecting 21 males for promotion was not due to chance variation and that there could be discrimination against women? Justify your answer.

The 1291 simulations shown in figure 4 represent a class data set compiled from two school terms. Figure 5 shows graphs of the results of the simulations from the two terms. The graph in figure 5a shows the results of 278 simulations conducted in one term, and that in figure 5b shows the results of 1013 simulations conducted in the other. Together, the graphs show a total of 1291 simulations. Answer questions 9–13 on the basis of the data in figure 5.

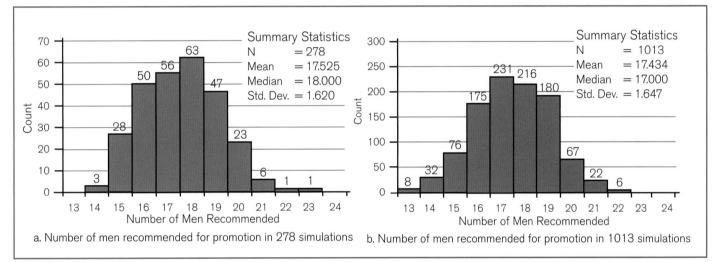

a. Number of men recommended for promotion in 278 simulations   b. Number of men recommended for promotion in 1013 simulations

Fig. 5. A breakdown of the results of 1291 simulations from two terms

9. *a.* Compare the two simulated sampling distributions with respect to shape, center, and spread.

b. For each term's simulated sampling distribution, estimate the probability that 21 or more of the 35 selected for promotion would be male. In figure 5a, 8 simulated counts were 21 or more. In figure 5b, 28 simulated counts were 21 or more.

10. On the basis of your answers to question 9, do you believe that the simulated sampling distributions would be consistent from one large group to the next?

Analyzing Simulation Results (continued)

Name _____

11. Compare your individual simulated sampling distribution of 20 simulated counts to these two simulated sampling distributions. Would your original analysis change or stay the same?

12. *a.* Use the collected class data set from the original task and compare this simulated sampling distribution to the simulated sampling distribution from the two large groups.

    b. Are you convinced that the results are repeatable?

13. Does the number of simulated counts for a distribution make a difference in the consistency of the results? Compare the distribution of your original 20 simulated counts to the 278 simulated counts. Compare the distribution of the 278 simulated counts to the distribution of the 1013 simulated counts.

Dietary Change and Cholesterol

High cholesterol is a contributor to heart disease. Table 1 lists data from a study investigating the effect of dietary change on cholesterol levels. Twenty-four hospital employees voluntarily switched from "a standard American diet" to a vegetarian diet for one month. The data show their cholesterol levels both before and after the dietary change, in milligrams of cholesterol per deciliter of blood (mg/dL). Suppose for the activities that it is always desirable to decrease the level of cholesterol in the blood. Thus, assume that the purpose of the switch to the new vegetarian diet was to decrease that level.

Table 1

Cholesterol Levels before and after Changing Diets

Before (mg/dL)	After (mg/dL)	Before (mg/dL)	After (mg/dL)
195	146	169	182
145	155	158	127
205	178	151	149
159	146	197	178
244	208	180	161
166	147	222	187
250	202	168	176
236	215	168	145
192	184	167	154
224	208	161	153
238	206	178	137
197	169	137	125

(From Rosner, Bernard, *Fundamentals of Biostatistics* [Boston: Duxbury Press, 1986])

Was the diet a success, or could decreases in cholesterol levels have been due merely to chance?

You will be exploring questions such as this in the following activities:

• What Can You Know—How Can You Show? (or the alternative activity Data-Based Dietary Decisions)

• Simulating and Counting Successes

What Can You Know— How Can You Show?

Name _____

Read (or reread) the scenario Dietary Change and Cholesterol completely before trying to answer questions 1–3. Then reinspect both the questions and the data in table 1 without attempting any data analysis. Then answer questions 1–3.

1. Suppose that a doctor were looking at the data. Write two or three questions that might be of interest in thinking about patients *and that could be answered by using only these data.*

2. Write two or three questions that could be answered by using these data and that you would want to answer if someone proposed that you change your diet to lower your cholesterol. Word your questions as precisely as you can.

3. For each question that you wrote in numbers 1 and 2, describe the kind of graph that you would make of the data to answer it. Specify the features of your graphs that you would examine, and tell how those features would help answer the questions.

4. Now use the data in table 1 to make the graphs that you described in number 3, and check the features that you identified.

Name _____

5. Answer the following questions on the basis of your analysis—that is, with information obtained directly from the data. (Do not be influenced by what you have learned or heard about cholesterol and diet elsewhere.)

   a. If you were asked to change from "a standard American diet" to a vegetarian diet to lower your cholesterol level, would you do it? Explain carefully.

   b. If you were a doctor or dietitian, would you recommend changing from "a standard American diet" to a vegetarian diet to lower the cholesterol levels of your patients or clients? Explain carefully.

Discussion and Extension

1. Think about how and from whom the data were collected. How widely do you think the conclusions of this study might be applied? What are your concerns?

Data-Based Dietary Decisions

Name _____

Read (or reread) the scenario Dietary Change and Cholesterol. Adopt the point of view of someone (doctor, dietitian, patient, etc.) who might be interested in using the results of this study.

1. What decisions would be appropriate for someone to ask you to make in your role as this person?

   Please list them.

2. Analyze, display, and interpret the data to justify these decisions or to call them into question.

Discussion and Extension

1. Suppose that the probability of cholesterol levels decreasing when people switched to a vegetarian diet really were 50 percent. If this were so, what would be the probability that at least 21 volunteers (out of 24) would improve in a study like the one here?

2. What would your answer suggest about the diet?

3. Think about how and from whom the data were collected. How widely do you think the conclusions of this study might be applied? What are your concerns?

Simulating and Counting Successes

Name _____

Observe that the cholesterol levels of three volunteers actually increased rather than decreased in the study described in the activity sheet "Dietary Change and Cholesterol." Could these "failures" mean that the change in diet was *not* successful in lowering cholesterol?

If the vegetarian diet actually had no effect, we would expect decreases and increases to occur purely randomly, with the two outcomes, "success" and "failure," being equally likely—that is, both having a probability of 1/2.

Suppose that the probability of cholesterol levels decreasing when people switched to a vegetarian diet really were 50 percent. Under that condition, what would be the probability that the cholesterol levels of at least 21 volunteers out of 24 would decrease in a study like that in the scenario Dietary Change and Cholesterol?

What would your answer suggest about the diet?

In this activity, you will use a simulation to estimate the chance of this outcome. For one complete replication of the simulation of random results for the 24 subjects in the cholesterol study—

- let a study volunteer's result be simulated by one toss of a fair coin;
- define heads to mean improvement (i.e., lowered cholesterol) and tails to mean failure to improve;
- toss the coin 24 times to simulate outcomes for the 24 volunteers in the study; and
- count and record the number of successes (heads) among the 24 trials (tosses).

1. Do the steps in the simulation represent the key elements of the situation? Explain.

2. Carry out 10 complete replications of the simulation. Record the number of successes separately for each simulation.

Simulation Number:	1	2	3	4	5	6	7	8	9	10
Number of Successes:										

Simulating and Counting Successes (continued)

Name _____

3. How many replications resulted in 21 or more successes? How likely do you think having at least 21 successes would be if the diet really did not work?

4. Combine your simulation data with those of your classmates. How many of all the replications resulted in 21 or more successes? How likely do you now think having at least 21 successes would be if the diet really did not work?

5. Construct a histogram of the data in number 4 and mark the region representing at least 21 successes. What conclusions can you draw?

What Does This Study Do?

Name _____

Read the following summaries of three studies of the effects of cell phone use. These studies attempted to investigate the possibility of a relationship between the use of cell phones and the development of cancer.

Study 1: In a study by Joshua E. Muscat and others reported in the *Journal of the American Medical Association,* the cell phone habits of 469 people with brain cancer were compared with those of 422 healthy people matched by age, gender, and other characteristics. The cell phone use of the group with cancer averaged 2.5 hours per month, compared with the control group's 2.2 hours per month. Although the difference of 0.3 hours on the cell phone each month was not significant, the average time on the cell phone of the group with cancer was higher.

> (From Muscat, Joshua E., Mark G. Malkin, Seth Thompson, Roy E. Shore, Steven D. Stellman, Don McRee, Alfred I. Neugut, and Ernst L. Wynder, "Handheld Cellular Telephone Use and Risk of Brain Cancer," *Journal of the American Medical Association* 284 [December 2000], pp. 3001–3007)

Study 2: In the London *Sunday Times,* Jonathan Leske reported on a study that investigated cell phone use as a cause of cancer. This study compared a group of 118 people with an eye cancer (uveal melanoma) with a control group of 475 people without the disease. The group with cancer had a much higher rate of cell phone use than the control group.

> (From Leske, Jonathan, "Scientists Link Eye Cancer to Mobile Phones," *Sunday Times* [London], 14 January 2001, News sec., p. 30 [www.newsint-archive.co.uk/pages/sample_pre.asp])

Study 3: Stewart Fist reported the results of an experiment investigating the link between cell phone use and tumors in mice. The study exposed one hundred mice to cell phone radiation for two half-hour periods each day over eighteen months. The researchers fitted another one hundred mice with the same type of antennas, which never had the power turned on. Tumor rate, the response variable measured, was twice as high in the exposed group as in the unexposed group.

> (From Fist, Stewart, "Cell Phones Cancer Connection," *Australian* [Sydney] 29 April 1997. [http://www.rense.com/health/cancer.htm])

What Does This Study Do? (continued)

Name _____

1. What is the question of interest in study 1? In study 2? In study 3?

2. Can you think of any problems that might have an impact on study 1? On study 2? On study 3?

3. Do you think that study 1 shows evidence that supports the conclusions? How about study 2? Study 3?

Discussion and Extension

1. Explain for each study whether you think that any results are applicable to all people who use cell phones.

Navigating through Data Analysis in Grades 9–12

How Fast Do They Melt in Your Mouth?

Name _____

A baker is trying to decide what chips to use in chocolate chip cookies that he plans to offer for sale. A salesman has told him that chocolate-flavored chips are "just like" real semisweet chocolate chips but cost much less. After thinking about it, the baker decides that what delights people most about chocolate chip cookies is the way the chocolate melts in their mouths. If the baker can be sure that there is no difference between real semisweet and chocolate-flavored chips in this respect, then he will use chocolate-flavored chips in his cookies.

Design an experiment to determine whether there is any difference in the average time that it takes for chocolate-flavored and real semisweet chocolate chips to melt in a person's mouth. To help in the design process, consider the following.

1. What will the treatment be?

2. What will the response variable be?

3. How will the response variable be measured?

4. What will the experimental unit be?

5. Specify the experimental conditions to be used by considering factors that might have an impact on the time that it takes a chip to melt in a person's mouth and what, if anything, should be done to control this variation.

6. What will the scope of inference be?

Name _____

7. Will it be possible to draw causal inference?

Discussion and Extension

Work together with the other students in your class.

1. Describe how this study could be conducted as a two-group experiment.

2. Describe how this study could be conducted using a paired design.

3. Conduct the study. If you use both paired and two-group designs, make graphs and use summary statistics to compare the results for the two types of chips in each study.

The Salk Polio Vaccine

Name _____

In 1954, the National Foundation for Infantile Paralysis (NFIP) and the U.S. Public Health Service were both preparing to conduct tests on the public at large of a vaccine developed by Jonas Salk. Infantile paralysis, also called poliomyelitis, is more commonly known as polio. The two groups approached the testing in different ways. Recognizing that children in grades 1–3 were most susceptible to polio, the NFIP focused its test on children in those grades. The NFIP study used children in grade 2 as the treatment group and gave them the vaccine. Children in grades 1 and 3 were used as the control group. All the children had to have parental consent to be vaccinated, however, and many of the second graders' parents did not give their consent (see table 1; note that the numbers are rounded).

Table 1

Polio Vaccine–NFIP Study

	The NFIP Study	
	Size	Rate of polio cases per 100,000
Grade 2 (vaccine)	225,000	25
Grades 1 and 3 (control)	725,000	54
Grade 2 (no consent)	125,000	44

(From Francis, Thomas, Jr., "An Evaluation of the 1954 Poliomyelitis Vaccine Trials–Summary Report," *American Journal of Public Health* 45 [1955])

Researchers in the Public Health Service used a different design. They selected a set of students whose parents had given consent and randomly assigned these children to the treatment (vaccine) or to the control group. All children who did not get the vaccine were given a placebo (a harmless injection of salt and water), so the children did not know whether they were given the treatment or not. In addition, the doctors who subsequently treated children with suspected polio did not know whether the children had received the vaccine or a placebo. Such an experiment is called *double-blind* because neither the subjects nor the evaluators know who has had the treatment. The results of this study are in table 2 (note that the numbers are rounded).

Table 2

Polio Vaccine–Public Health Study

	Randomized Controlled Double-Blind Experiment	
	Size	Rate of polio cases per 100,000
Treatment	200,000	28
Control	200,000	71
No consent	350,000	46

(From Francis 1955)

The Salk Polio Vaccine (continued)

Name _____

Reflect and analyze

1. Comment on the two methods with respect to the question of interest, the design of the studies, the role of randomization, and the results.

2. Show how two-way tables can help in analyzing the problem.

3. Contrast the populations being sampled in the two cases. Could the differences between the populations explain the difference in the outcomes of the two studies?

Medical Studies

Name _____

Study 1: Sleep apnea may be linked to Alzheimer's

Hiroshi Kadotani and others (2001) reported in the *Journal of the American Medical Association* that "a disorder that causes breathing lapses during sleep is linked to a gene variation associated with Alzheimer's and cardiovascular problems." The researchers wrote that sleep-disordered breathing could be a result of a gene variation called apolipoprotein E-4. Sleep apnea, characterized by snoring and brief breathing lapses during sleep, is mostly hereditary. In the study, the risk of sleep apnea for participants with the E-4 gene variation was almost double the risk of sleep apnea for those without the trait. The sleep disorders study involved 791 middle-aged adults, ranging from 32 to 68 years of age. Of the 222 participants with the E-4 variation, 27 had moderate to severe apnea, as compared with 39 of the 569 participants without the variation. According to the study, a significant portion of sleep-disordered breathing is associated with the gene variation in the general population. The authors said that sleep apnea may interact with the E-4 trait to impair mental abilities, but their study does not necessarily imply that all adults with sleep apnea face an increased risk of Alzheimer's or cardiovascular disease.

(From Kadotani, Hiroshi, Tomiko Kadotani, Terry Young, Paul E. Peppard, Laurel Finn, Ian M. Colrain, Greer M. Murphy, and Emmanuel Mignot, "Association between Apolipoprotein E-4 and Sleep-Disordered Breathing in Adults." *Journal of the American Medical Association.* 285 [June 2001], pp. 2888–2890)

Study 2: Tirofiban improves the success rate of bypass surgery

Gilles Montalescot and others (2001) found that the heart drug tirofiban helped patients who were undergoing bypass surgery. This study randomly assigned 300 patients in a double-blind fashion to either tirofiban (149 patients) or a placebo (151 patients) before they underwent coronary angiography, a procedure to expand the artery. After thirty days, death or major setbacks had occurred in 9 of those in the tirofiban group, as compared with 32 of those in the placebo group. The investigators concluded that early administration of tirofiban improved the success rate of the bypass surgery.

(From Montalescot, Gilles, Paul Barragan, Olivier Wittenberg, Patrick Ecollan, Simon Elhadad, Philippe Villain, Jean-Marc Boulenc, Marie-Claude Morice, Luc Maillard, Michel Pansieri, Remi Choussat, and Philippe Pinton, "Platelet Glycoprotein IIb/IIIa Inhibition with Coronary Stenting for Acute Myocardial Infarction," *New England Journal of Medicine,* 344 (June 2001), pp. 1895–1903)

Medical Studies (continued)

Name _____

Reflect and analyze

Discuss the two summaries with respect to the following points:

1. Were the investigations observational studies or experiments?

2. What were the questions of interest?

3. What were the designs of the studies?

4. What role did randomization play in the studies?

5. What conclusions could be drawn from the studies?

Darwin and Genetics

Name _____

Charles Darwin was interested in whether there was any difference in the height of plants that were cross-fertilized—that is, that were fertilized by pollen from other, closely related plants—and those that were self-fertilized. The data in table 1 show the height in inches of plants grown in 15 pairs for a fixed period of time. One member of each pair was cross-fertilized, and the other was self-fertilized.

Table 1.

Plant Height

Pair	Cross-Fertilized (in.)	Self-Fertilized (in.)
1	23.5	17.4
2	12.0	20.4
3	21.0	20.0
4	22.0	20.0
5	19.1	18.4
6	21.5	18.6
7	22.1	18.6
8	20.4	15.3
9	18.3	16.5
10	21.6	18.0
11	23.3	16.3
12	21.0	18.0
13	22.1	12.8
14	23.0	15.5
15	12.0	18.0

(From Darwin, Charles, *The Effect of Cross- and Self-Fertilization in the Vegetable Kingdom.* London: John Murray, 1876)

Reflect and analyze

What conclusions can you draw from these data? Justify your conclusions by using techniques discussed in your class.

Examining an Article in the News Media

Name _____

Look on the Internet or in the newspaper and find a news article describing a study. Write a description of the study. Include the following components in your description:

- The question under investigation

- The design of the study

- The nature of the data collected

- Any assumptions made by the researchers

- The role of randomization

- The results

- Your interpretation of the conclusions

Designing a Study

Name _____

Design a study to explore a question of particular interest to you. Be sure that you state the question, any assumptions that you will make, what data are important and how you will collect them, the role of randomization, and how you propose to use probability in drawing your conclusions.

Some possible topics are suggested:

- Does studying with the television on make a difference in students' grade-point averages?

- Does regular exercise lower pulse rate?

- Is there a difference between two brands of microwave popcorn in the amounts popped?

- What characteristics of a paper airplane enable it to fly farther?

Solutions for the Blackline Masters

Solutions for "Sampling Rectangles"

1–4.   Solutions will vary and are described in the text.

Solutions for "Sample Size"

1–3.   Solutions will vary and are described in the text.

Solutions for "Sampling Methods"

Task A—Stratified Samples
1–3.   Solutions will vary and are described in the text.

Task B—Clusters Samples
1–4.   Solutions will vary and are described in the text.

Solutions for "What Would You Expect?"

1–4.   Solutions are described in the text.

5.   *a.* 60 percent
      b. 40 percent

6.   Students' answers will vary. The study data are in the "gray" area.

7.   Students might suggest that 50 percent of those recommended for promotion would be female and 50 percent of those recommended for promotion would be male, since the numbers of male and female applicants were equal (24 of each). Thus, they may (or may not) conclude that 60 percent to 40 percent would not be that "far off." It is important to note that if the numbers of male and female applicants were not equal, then we would not compare the observed percentages of applicants recommended for promotion to 50 percent for males and 50 percent for females. For example, if there were 30 male applicants and 18 female applicants and no discrimination, we would expect to observe that of the applicants recommended for promotion, 22 would be male (about 63 percent) and 13 would be female (about 37 percent).

Discussion and Extension

1.   Suggestions may include looking at the past history of the bank and the supervisors, using probability to see how likely the results are to occur, or interviewing the supervisors.

2.   Students may raise considerations related to sample size (20 of each may be considered small), or they might ask whether the supervisors were male or female, or how supervisors had promoted males and females in the past. Students should begin to think about the design of the study.

Solutions for "Simulating the Case"

1–8. Solutions will vary depending on the students' responses to the activity. Students' responses will depend on their simulations and should take the form described in the discussion for figure 2.1. Students will probably suggest that the distributions from their simulations are close to what they expected. Teachers might push to get them to describe exactly what they expected to see.

Discussion and Extension

1. In most cases, students' simulations will show that the probability of recommending 21 males of the 35 candidates recommended for promotion is approximately 0–10 percent.

2. Typically, the students' results will be similar to one another, although occasionally a simulated sampling distribution will be very different from that expected.

Solutions for "Analyzing Simulation Results"

1–10. Solutions will vary and are described in the text.

11. Answers will depend on the students' individual simulated sampling distributions. Using the simulated sampling distributions given in the text as an example, we see that the shape of the simulated sampling distribution of 20 counts is approximately symmetrical and close to what is expected, though slightly skewed to the right. The sample median is 17.5, and the sample mean is 17.7. The counts vary from 13 to 23; however, most of the data fall between the values of 16 and 19. The sample interquartile range is 2, from 16.5 to 18.5. The standard sample deviation is 1.9 units.

12. Answers will depend on the students' individual simulated sample distributions. Using the distributions given in the text as an example, however, we see the following:

   a. The mean based on the simulated results was close to the expected mean of 17.5.

   b. A description is given in the discussion in the text.

13. The solution is described in the text.

Solutions for "What Can You Know—How Can You Show?"

1. Sample questions include the following:
   a. Does the diet make a difference?
   b. Does the diet work (or not work) for a given set of patients?
   c. Can I predict the change in a patient's cholesterol level if he or she uses the diet?

2. Sample questions include the following:
   a. About how much can I expect my cholesterol level to drop?
   b. What is the largest decrease that I can reasonably expect if I change diets?
   c. What are the chances that my cholesterol level will not drop?
   d. How does my current cholesterol level affect the answers to the previous questions?

3. Note that there could be different answers for each. The answers given correspond to the sample answers provided above to (1) and (2).
   Appropriate graphs answering the sample questions in (1) include, respectively—

   a. A scatterplot of *After* versus *Before*
   b. A box plot of improvement
   c. A dot plot of improvement

   Appropriate graphs answering the sample questions in (2) include, respectively—

   a. A scatterplot with a regression line
   b. A dot plot of the differences *After − Before* (or *Improvement*).
   c. A dot plot of change and a calculation of the probability.
   d. A scatterplot of *After* versus *Before.*

In response to the question about what feature of the graphs they might examine, students might suggest that they would—

- look for the center of the distribution of differences in cholesterol level;
- find the maximum value (right endpoint) of the distribution of differences in cholesterol level;
- compute the ratio of the number of cholesterol differences below 0 to the total number of observations (24).

Looking at question (*d*) in (2), for example (How does my current cholesterol level affect the answers to the previous questions?), students might give responses like the following to explain how specific features would help them answer the question:

- I could find my current cholesterol value along the horizontal axis and then estimate the center of the levels with similar "before" cholesterol values (the least squares line provides a reasonable estimate).
- I could find my current cholesterol value along the horizontal axis and then estimate the lowest "after" cholesterol value from the dots with similar "before" cholesterol values.
- I could find my current cholesterol value along the horizontal axis and then compute the ratio of the number of points above the line *After = Before* to the number of all points with similar "before" cholesterol values.

4. See sample solutions discussed in the chapter.
5. *a.* A sample answer follows: "If I really wanted to decrease my cholesterol level and my initial level were at least 175, yes. However, since many doctors view levels below 200 as satisfactory, I might use that as my threshold for changing diets. Also, taking into account the way in which the data for this study were generated, I might want to wait for more information!"
   b. A blanket recommendation to change would appear to present little danger. However, patients starting below 175 have a reasonable chance of failing to improve. If initial cholesterol levels are available, more targeted recommendations (see sample answer to 5*a*) would probably be received more favorably.

Discussion and Extension

1. Sample answers follow:
   - I don't think they apply to many people beyond those who took part.
   - The sample is too small (only 24 people).
   - The sample is only from hospital employees (they may be healthier than or eat different types of food from the general population, etc.)
   - The sample is made up only of volunteers.

Solutions for "Data-Based Dietary Decisions"

This activity is open-ended, and answers will vary. Answers could be expected to be similar to those for What Can You Know—How Can You Show?

Solutions for "Simulating and Counting Successes"

This activity is a simulation, and all answers will depend on the simulation.

Solutions for "What Does This Study Do?"

1. *Study 1:* The question was whether there is a link between cell phone use and brain tumors.
   Study 2: The question was whether there is a connection between cell phone use and a particular eye cancer.
   Study 3: The question was whether cell phone radiation caused tumors in mice.

2. *Study 1:* We do not know how the people were chosen. Were there other factors besides cell phone use that the people with eye cancer had in common that could explain their cancer?

   Study 2: We do not know how the people were chosen. Were there other factors besides cell phone use that the people with eye cancer had in common that could explain their cancer?

   Study 3: Do mice develop cancer in the same way that humans do?

3. Answers will vary for each study. Because of students' limited knowledge, you should expect them to base more of their discussion on experience than on data analysis.

Discussion and Extension

1. *Study 1:* The results are probably applicable only to the people who were involved in the study.

   Study 2: The results are probably applicable only to the people involved in the study.

   Study 3: The results may be applicable only to the mice involved in the study.

Solutions for "How Fast Do They Melt in Your Mouth?"

Answers to this activity will depend on the students. We recommend using this activity as the basis for a class discussion after students have had time to think about the design. A complete discussion of the activity appears in the text.

Solutions for "The Salk Polio Vaccine"

1–3. Solutions are described in the text.

Solutions for "Medical Studies"

1–5. Solutions are described in the text.

Solutions for "Darwin and Genetics"

Solutions are described in the text.

References

Barbella, Peter, Lorraine Denby, and James M. Landwehr. "Beyond Exploratory Data Analysis: The Randomization Test." *Mathematics Teacher* 83 (February 1990): 144–49.

Barbella, Peter, James Kepner, and Richard Scheaffer. *Exploring Measurements.* Palo Alto, Calif.: Dale Seymour Publications, 1994.

Bright, George W., Wallece Brewer, Kay McClain, and Edward S. Mooney. *Navigating through Data Analysis in Grades 6–8. Principles and Standards for School Mathematics* Navigations Series. Reston Va.: National Council of Teachers of Mathematics, 2003.

Burke, Maurice, David Erickson, Johnny W. Lott, and Mindy Obert. *Navigating through Algebra in Grades 9–12. Principles and Standards for School Mathematics* Navigations Series. Reston Va.: National Council of Teachers of Mathematics, 2001.

Cobb, George W., with Jonathan D. Cryer. *An Electronic Companion to Statistics.* New York: Cogito Learning Media, 1997.

Darwin, Charles. *The Effect of Cross- and Self-Fertilization in the Vegetable Kingdom.* London: John Murray, 1876.

Errthum, Emily, Maria Mastromatteo, Vince O'Connor, and Richard Scheaffer. *Exploring Projects.* White Plains, N.Y.: Dale Seymour Publications, 1999.

Fist, Stewart. "Cell Phones Cancer Connection," *Australian* (Sydney), 29 April 1997. (http://www.rense.com/health/cancer.htm)

Francis, Thomas, Jr. "An Evaluation of the 1954 Poliomyelitis Vaccine Trials—Summary Report." *American Journal of Public Health* 45 (1955): 1–63.

Freedman, David, Robert Pisani, and Roger Purves. *Statistics.* 3rd ed. New York: W. W. Norton & Company, 1998.

Good, Phillip I. *Resampling Methods.* Boston: Birkhäuser, 1999.

Huff, Darrell. *How to Lie with Statistics.* New York: W. W. Norton & Company, 1954.

Kadotani, Hiroshi, Tomiko Kadotani, Terry Young, Paul E. Peppard, Laurel Finn, Ian M. Colrain, Greer M. Murphy, and Emmanuel Mignot. "Association between Apolipoprotein E-4 and Sleep-Disordered Breathing in Adults." *Journal of the American Medical Association* 285 (June 2001): 2888–90.

KCP Technologies. Fathom. Emeryville, Calif.: Author, 2000.

Leske, Jonathan. "Scientists Link Eye Cancer to Mobile Phones," *Sunday Times* (London), 14 January 2001, News sec., p. 30. (www.sundaytimes.co.uk/news/pages/sti/2001/01/14/stinwenws01032.html)

Montalescot, Gilles, Paul Barragan, Olivier Wittenberg, Patrick Ecollan, Simon Elhadad, Philippe Villain, Jean-Marc Boulenc, Marie-Claude Morice, Luc Maillard, Michel Pansieri, Remi Choussat, and Philippe Pinton. "Platelet Glycoprotein IIb/IIIa Inhibition with Coronary Stenting for Acute Myocardial Infarction." *New England Journal of Medicine* 344 (June 2001): 1895–1903.

Moore, David. *The Basic Practice of Statistics.* New York: W. H. Freeman and Company, 2000.

Muscat, Joshua E., Mark G. Malkin, Seth Thompson, Roy E. Shore, Steven D. Stellman, Don McRee, Alfred I. Neugut, and Ernst L. Wynder. "Handheld Cellular Telephone Use and Risk of Brain Cancer." *Journal of the American Medical Association* 284 (December 2000): 3001–3007.

National Council of Teachers of Mathematics (NCTM). *Principles and Standards for School Mathematics*. Reston VA: NCTM, 2000.

Ramsey, Fred L., and Daniel W. Schafer. *The Statistical Sleuth: A Course in Methods of Data Analysis*. Belmont, Calif.: Duxbury Press, 1996.

Repacholi, Michael H., Antony Basten, Val Gebski, Denise Noonan, John Finnie, and Alan W. Harris. "Lymphomas in Eμ-*Pim 1* Transgenic Mice Exposed to Pulsed 900 MHz Electromagnetic Fields," *Radiation Research* 147 (May 1997): 631-40.

Rosen, Benson, and Thomas H. Jerdee. "Influence of Sex-Role Stereotypes on Personnel Decisions," *Journal of Applied Psychology* 59 (February 1974): 9–14.

Rosner, Bernard. *Fundamentals of Biostatistics*. Boston: Duxbury Press, 1986.

Suggested Reading

Barrett, Gloria B. "Investigating Distributions of Sample Means on the Graphing Calculator." *Mathematics Teacher* 92 (November 1999): 744–47.

Edgerton, Richard T. "Analyzing Data from the Olympic Games for Trends and Inferences." *Mathematics Teacher* 89 (May 1996): 370–72.

Iversen, Gudmund R., and Mary Gergen. *Statistics: The Conceptual Approach*. New York: Springer-Verlag, 1997.

McClintock, Edwin, and Zhonghong Jiang. "Spreadsheets: Powerful Tools for Probability Simulations." *Mathematics Teacher* 90 (October 1997): 572–79.

Moore, David. *Statistics: Concepts and Controversies*. New York: W. H. Freeman and Company, 1997.

Ott, Lyman, Richard F. Larson, and William Mendenhall. *Statistics: A Tool for the Social Sciences*. 4th ed. Boston: PWS-Kent Publishing Company, 1987.

Piccolino, Anthony V. "The Advanced Placement Course in Statistics: Increasing Students' Options." *Mathematics Teacher* 89 (May 1996): 376–77.

Scheaffer, Richard L., and James T. McClave. *Probability and Statistics for Engineers*. 3rd ed. Boston: PWS-Kent Publishing Company, 1990.

Scheaffer, Richard L., William Mendenhall, and Lyman Ott. *Elementary Survey Sampling*. Boston: POWS-Kent Publishing Company, 1990.

Shulte, Albert P., and Jim Swift. "Data Fitting without Formulas," *Mathematics Teacher* 79 (April 1986): 264–71, 297.

Vonder Embse, Charles, and Arne Engebretsen. "Visual Representations of Mean and Standard Deviation," *Mathematics Teacher* 89 (November 1996): 688–92.

Walton, Karen Doyle. "Probability, Computer Simulation, and Mathematics." *Mathematics Teacher* 83 (January 1990): 22–25.